Feb 9, 2008
Sarah,
Best Wishes
on your Birthday.

[signature]

HOW TO
Think
LIKE A
Horse

HOW TO
Think
LIKE A
Horse

The Essential Handbook for Understanding
Why Horses Do What They Do

CHERRY HILL

 Storey Publishing

The mission of Storey Publishing is to serve our customers by publishing practical information that encourages personal independence in harmony with the environment.

Edited by Deborah Burns

Art direction by Vicky Vaughn

Text design and production by Jennie Jepson Smith

Cover photograph by © Cherry Hill: Randy Dunn's horses at Bath Brothers Ranches

Interior photographs by © Mark J. Barrett: 132; © Royalty-free/CORBIS: chapter headings (background texture), 73; © Cherry Hill: vi 3rd and 4th from top, vii bottom, 6, 8, 20, 26, 33, 37, 48, 58, 64, 65, 81, 116, 122, 143, 151, 158, 160 right; © Richard Klimesh: vi top two, vii 3rd from top, 3, 4, 12, 13, 15, 16, 18, 23, 24, 30, 43, 46, 57, 67, 68, 75, 80, 91, 124, 131, 133, 142, 144, 145, 146, 147, 148, 152, 153, 160 left, 161 left, 167; © Bob Langrish: ii, v, vi bottom two, vii top two, x, 5, 14, 22, 29, 31, 32, 34, 60, 62, 69, 74, 77, 79, 98, 104, 111, 117, 124, 135, 140, 141, 161 right; © Storey Publishing: 40–41

Illustrations by © Elayne Sears

Image coordination by Ilona Sherratt

Image management by Laurie Figary

Prepress by Kevin Metcalfe

Infographics by Kristy MacWilliams: 9, 72; Ilona Sherratt: 36, 44, 128, 174

Indexed by Susan Olason, Indexes & Knowledge Maps

Special thanks to Country Tack, Lanesborough, Mass., for grooming tools shown on pages 40–41.

The information in this book is true and complete to the best of our knowledge. All recommendations are made without guarantee on the part of the author or Storey Publishing. The author and publisher disclaim any liability in connection with the use of this information. For additional information please contact Storey Publishing, 210 MASS MoCA Way, North Adams, MA 01247.

Storey books are available for special premium and promotional uses and for customized editions. For further information, please call 1-800-793-9396.

Printed in China by Regent Publishing Services

10 9 8 7 6

LIBRARY OF CONGRESS CATALOGING-IN-PUBLICATION DATA

Hill, Cherry, 1947–
  How to think like a horse / by Cherry Hill.
    p. cm.
  Includes index.
  ISBN-13: 978-1-58017-835-8 (pb : alk. paper)
  ISBN-13: 978-1-58017-836-5 (hc : alk. paper)
  1. Horses—Behavior. 2. Human-animal communication. 3. Horsemanship. I. Title.
SF281.H55 2006
636.1—dc22
                                                  2005027792

To my long-time friends:

Deborah Burns, editor extraordinaire for 20 years

*Sassy Eclipse — "Sassy," excellent trail horse and
broodmare for 30 years*

*Miss Debbie Hill — "Zinger," generous dressage and
western horse for 30 years*

and especially to

*Richard Klimesh, my husband and best friend
for over 35 years.*

Contents

Preface

WHEN I WAS A VERY YOUNG CHILD, I not only wanted to be with horses all of the time, but I even wanted to *be* a horse. I galloped, reared, kicked, and nickered. When I saw a new thing, I'd walk up very cautiously, roll my head forward and down to get a really good look, and then I'd jump lightly to the side with a squeal. Then I'd approach the item again to smell it with an air of suspicion and high alertness, all the while making snorting and blowing sounds.

I even did this at the dinner table to inspect my food. Not the greatest behavior when we had company, especially when I'd follow the sniffing with a loud whinny. Our guests would joke to my parents that I was part horse, but that was not good enough for me. I wanted to be all horse. That's why, after grooming a horse, I'd be careful not to completely wash my hands so that I could save that wonderful smell for as long as possible.

Somewhere in grade school, much to my parents' relief, my external horse behavior subsided somewhat, but the core of my being had become part horse. And so it is to this day. My life has revolved around horses.

Many of the books I have written about horses have been about horse-related "things" — barns and feeding, grooming and tack, hoof care, training and riding. In a few books, I've talked about the interaction between humans and horses as it relates to management or training, but up until now, I haven't devoted an entire book to what makes up the horse.

Since I am a teacher at heart, my explanations have lots of facts and details in them. This is because instead of having you take my word for it, I'd rather give you something you can sink your teeth into — information that can help you come to your own conclusions. Because research on horse behavior is limited and opinions abound, I'm offering you my interpretation of what has been published.

Yet, when it comes right down to it, all of the books and information in the world are not going to take the place of the time and experience needed to develop a sense of horse. As far as I am concerned, that is something very personal. No one can give it to you, you can't buy it, and you can't come to it purely from an intellectual standpoint. Learning some facts about horses, however, will help you form a valuable base for understanding their needs, behaviors, and abilities. To that end, I have written a combination "left brain/right brain" book directed at the scientist and artist that exist in all of us.

Besides the facts, I'm including examples and anecdotes that will bring those facts to life. Throughout, you'll read my interpretation of the facts based on my lifetime with horses as a trainer, riding instructor, classroom teacher, horse show judge, breeder, book author, and photojournalist.

Nevertheless, I encourage you to be open-minded. Listen to, watch, and read other horsemen. Let the subterfuges sink, and watch the cream rise to the top. Skim off what you want to help you develop your own sense of what is horse.

Traditionally, many terms describing people that work with horses have been masculine — *horseman* and *horsemanship*. Yet today, women make up the majority of horse owners. And what is even more interesting is that many of the so-called newer methods of natural horse training, advocated primarily by male clinicians as being less harsh and more of a thinking type of training, are how women instinctively approach animal training, anyway. Since we are the weaker sex physically, we tend to use our brains to avoid confrontations. We try to figure things out to keep from getting hurt. Maybe we are very similar to horses in that we both have a strong self-preservation aptitude. We also tend to be able to break things into small tasks and appreciate small efforts and partial progress.

Now I'm not saying that all male horse trainers are impatient and use physical means of domination or are bronc-busting brutes —

far from that! There have been and are many talented and thoughtful horsemen. But it can be a male tendency to be ruled by ego and testosterone, which can give rise to fighting and dominance. So although I'm delighted that there has been a shift in thinking about horse training, in the back of my mind I chuckle and think that we girls have known these things, inherently, all along.

In this book, I often use the term *horse trainer* to refer to the human because I figure that no matter how well trained a horse already is and no matter what we are doing with him, we are always training — whether adding new behaviors, modifying an existing behavior, or reinforcing an already established behavior. It doesn't matter if you are feeding a horse, if you are riding at a lope across a pasture, or if your farrier is shoeing your horse: the horse is learning and is being trained, whether or not you think of it as a formal training session.

I hope this book gives you a good idea of how and why a horse does what he does. This knowledge can give you a start in how to read body language and behave around horses, and help you plan and conduct your training sessions. For specific how-to training instructions, please refer to the recommended reading guide. Horse-training techniques require separate volumes.

Cherry Hill

CHAPTER

1

Becoming the Horse

WHEN MOST OF US LOOK AT A HORSE, we can easily see his beauty and admire his nobility. It's when we start interacting with him that things can go wrong. That's because we tend to view horses in human terms.

When a horse runs away from us, bucks, or bites, for example, we are likely to interpret what we think happened rather than observing and reporting specific objective facts. A human interpretation might be "That horse doesn't like me, is misbehaving, or is mean." Once a person comes to understand horses, however, such natural behaviors are seen for just what they are and actually become less frequent. The reason is that the better you understand horses, the less often conflict will arise.

Do horses think? Well, that depends on the definition. If thinking is using the mind to process information received from the senses, then of course horses think. But do they reason? If reasoning is using logic to come to a conclusion, then horses generally do not reason. Instead, they observe, react (often very quickly), and think later.

Why Think Like a Horse?

There are probably as many answers to that question as there are horsemen, and most people would cite a blend of many reasons. Here are some of the most common answers.

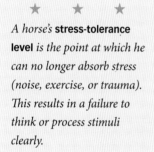

A horse's **stress-tolerance level** *is the point at which he can no longer absorb stress (noise, exercise, or trauma). This results in a failure to think or process stimuli clearly.*

   Instinct *is inborn, intrinsic knowledge and behavior.*

★ **To understand the world** from the horse's viewpoint.

★ **To make a horse feel as relaxed** around you as if he were with another horse.

★ **To communicate** with a horse in terms he can understand in order to persuade him to do what you ask. Horses are very willing, cooperative animals, so if what you ask is fair and possible and the horse understands you and is relaxed, you will have a better chance for success.

★ **To be safe.** Accidents are often a result of a misunderstanding. The more you can think like a horse, the less likely a horse will be to panic or have an explosive reaction.

★ **To have a satisfying, smooth experience.** When things are going wrong between a horse and a human, everything is awkward and out of sync. When things are going right, it is like a dance with perfect timing and grace.

★ **To minimize stress.** A little bit of stress is good — as our moms told us, it builds strong character. But let's face it: both you and your horse would rather have your relationship be low stress, comfortable, and harmonious. If you are both on the same wavelength, it can be. When practicing tai chi, two of the goals are learning to recognize when your hackles are up and developing a means to smooth them down. So it goes with handling horses. We need to identify when we are part of the problem and then learn how to become part of the solution.

★ **To achieve goals.** The more you can think like a horse, the more you will be able to communicate like a horse, and the faster you and he will progress. Most often with horses, I feel that the slower you go, the faster you'll get there.

★ **To help a horse become solid and confident.** The more you work with his natural behavior and instincts, the more thorough and long lasting the results will be.

★ **To have a win-win situation.** For you to succeed, it is not necessary for the horse to lose. You can both develop as you work together and emerge as winning individuals *and* a winning team.

★ **To add to the horse,** not take away from him. It is not necessary to break a horse into fragments; rather, if you know horses, you will be able to add to the horse and help him develop to his full potential.

★ **To help you get in touch** with your animal sense and become a better person. Working with animals can bring great rewards on many levels — physical, emotional, intellectual, and spiritual. You can become more compassionate, more physically fit, and more observant.

But the best reason of all for you to become part horse and think like a horse is that your actions will help preserve your horse's spirit. After all, that is what attracted us to horses in the first place.

If you think like a horse it will be easier to form a partnership.

The Spirit Lives On

In 1973, after I graduated from college and had the opportunity to train some horses and observe other horse trainers, I saw that there was something missing in conventional horse training. It seemed to me that most of the procedures relied on force and domination, and training was on a thirty-day timetable. I hoped to do better by horses by focusing on what each individual horse needed.

When I opened my first training business, which I called The Spirit Lives On, I offered my services at two for the price of one. My monthly fee was the same as that of other local trainers, but when someone brought me a horse to work with for thirty days, I'd take him on two conditions. First, the horse had to stay for at least sixty days. Although they were billed for sixty days of board, they were charged for only thirty days of training. And second, before I turned over the reins, the owner had to come and work with me and the horse for the last week or two before the horse went home.

I put in a lot of extra work during those early years, and I think it was appreciated by all of the horses and most of the owners (as always, some customers are just in a hurry). Like many young trainers, almost all of the horses I initially received for training were ones that had been started but had developed bad habits. It takes much more time to sort out undesirable habits than it does to start a horse properly from scratch. What's worse, many problem horses have damaged spirits; some that I trained were mentally rattled, agitated, bruised, or worn thin and ragged.

It was then that I learned by looking (but not peering) into a horse's eyes, down into his soul, that once the light had grown dim or gone out, rekindling it was difficult to impossible. I vowed never to cause a horse to tune out. And I wanted to know how and why horses withdrew.

My role gradually began to broaden into that of an instructor and educator. I wanted to teach people in the hopes that I could make some things better for some horses.

What a Horse Needs

Much equine behavior stems from instincts developed over millions of years of living in a nomadic herd.

If you know what horses like, want, and need, and what they dislike, don't want, and do not need, you will be more able to think like a horse. I'll start with a brief overview, because all of these topics will be discussed in more detail later in the book.

Am I talking about wild or domestic horses? Although most of the horses we handle today are born into domestication, wild horse instincts still form the basis for their behavior. Domestic horses have the same needs, fears, and innate patterns as their wild ancestors did, and their physical makeup hasn't changed much in the last several million years.

A horse's list of needs and wants might be ranked something like this:

1. Self-preservation — avoiding being injured or eaten by a predator
2. Eating and drinking for survival
3. Procreation
4. Socialization and routines

Self-Preservation

As a prey animal, the horse has survived by being wary of predators, which include the dog and cat families and humans. That's why horses are alert, wary, and suspicious, have a highly developed flight reflex, and will fight when threatened.

They don't like to be chased or cornered. They are social creatures that find safety in numbers. They have learned where to go and where not to go; which sights, sounds, and smells mean danger; where food and water are; and how to escape when danger is imminent. Although man is the ultimate predator (don't get me started), horses can learn to overcome their strong instincts of self-preservation and trust us.

Wild horses seek shelter from weather extremes and insects. While domestic horses should be provided with a safe, comfortable place to live, they do not want or need to be locked in a stall at the first raindrop or snowflake. Often, horses choose to stand out in the open rather than in the confines of a stall or pen.

Many of the subjects later in this book, such as a horse's senses, reflexes, and behavior patterns, will tie in with the survival imperative of self-preservation.

The Need for Feed

Don't kid yourself. Sure, your horse loves you, but when it comes right down to it, eating is much more important than being petted. In the wild, horses eat for 12 to 16 hours every day, ingesting the dry equivalent of 25 to 30 pounds of natural feed each day. (For purposes of discussion, natural feed is native pasture or grass hay.) But wild horses are constantly on the move. If allowed, domestic horses would eat 16 hours a day, too, but they do not need that much. Without feeding management, domestic horses would eat themselves sick, especially if what they are eating includes grain or alfalfa hay.

You can limit their intake, yet horses still have a strong urge to chew for many hours of the day. This need to chew can be satiated by feeding long-stem hay. I usually have four or more types of hay in my barn. One of them is "busy hay," which is mature grass hay, high in roughage and low in protein and energy. It comes in handy to use as the "satisfier" part of the ration. If adequate roughage is not supplied, horses may eat bedding or chew wood or the manes and tails of other horses.

When I first approached this foal on a Wyoming range, he was ready to flee: two legs in motion, head up, tail raised.

I quit advancing and stood still, and almost immediately he became relaxed and curious: lowered head reaching forward, soft tail, and all four feet on the ground.

I recommend feeding grass hay at least three times a day at the rate of 1.5 to 1.75 times the horse's body weight per day. For a 1,000-pound horse, that would be 15 to 17.5 pounds of hay per day, divided into three or more feedings. If a mixed-grass pasture is available, you can substitute pasture for some of the hay, as long as grazing is monitored carefully. Avoid alfalfa, and feed grain only if required for growth, breeding, or hard work. I feed very little grain, even to weanlings and yearlings.

How Often Should a Horse Eat?

To see firsthand how your horse reacts to various feeding frequencies, take five weeks to run this test.

Week 1: Feed him his entire ration once a day.

Week 2: Feed him twice a day at twelve-hour intervals, such as at 6:00 A.M. and 6:00 P.M.

Week 3: Feed him three times a day at regular eight-hour intervals, such as at 6:00 A.M., 2:00 P.M., and 10:00 P.M.

Week 4: Feed him four times a day at five- to six-hour intervals, such as at 6:00 A.M., noon, 5:00 P.M., and 10:00 P.M.

Week 5: Feed him free-choice grass hay. Allow him to have access to it for sixteen hours out of each day, such as between 10:00 P.M. and 8:00 A.M., and between noon and 6:00 P.M.

In each instance, note the following:

1 Did he finish all the feed in one eating session?

Except for the free-choice option, a horse should clean up all he is fed within two hours. If there is feed left, it will often be trampled, fouled, and wasted.

2 How long did he take to eat?

Ideally, a horse will eat his ration in one to two hours.

3 Did he leave any feed? Was it wasted, or did he come back and finish it later?

If there is feed left at the next feeding, the feed is bad or the horse is being fed too much. If a horse eats half his ration, walks over to get a drink, rests for five minutes, and then resumes eating, that is perfectly normal.

4 How eager (noisy, pawing, pushy) was he to be fed the next time he saw you?

A horse that is aggressive at feeding time is either being fed too little, too infrequently, or considers you below him on the pecking order.

5 Rate your horse's overall contentment on a scale of 1 to 10 for each feeding frequency.

Because eating is one of your horse's top priorities, his behavior at feeding time and overall contentment speak volumes. The method with the highest score will tell you which feeding frequency your horse prefers.

Now you have a better idea not only of what your horse considers important but also how he reacts to changes in his routines.

Healthy Grazing

Grazing management of domestic horses is a delicate balancing act among what a horse wants, what he needs, and what is best for the land. (See *Horsekeeping on a Small Acreage*, second edition, for detailed recommendations on both horse and pasture management.)

If we were to let our horses graze free choice, in many cases it would result in overgrazed land and overweight horses. It is necessary to monitor pasture growth and manage grazing to fit the carrying capacity of the land and the nutritional needs and health of the horse.

Horses do well on native pastures and lower-quality improved pastures, but rich pastures and alfalfa fields can lead to obesity, colic, and laminitis. We might be surprised when horses eagerly eat certain non-grassy weeds, such as fuzzy dandelion and prickly thistle. They tend to avoid poisonous plants unless there is nothing else to eat and they are hungry. A horse's inherent wisdom and keen senses of smell and taste usually help him discriminate between what is healthy and what is not.

Clean, Accessible Water

Horses drink 5 to 10 gallons of water a day — more in summer, often less in winter. They seek water about an hour after ingesting the majority of their roughage.

The grass is always greener on the other side.

During the summer a horse might drink three or four times per day.

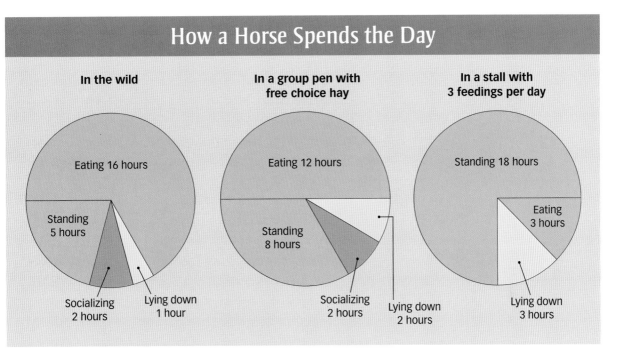

How a Horse Spends the Day

In the wild
- Eating 16 hours
- Standing 5 hours
- Socializing 2 hours
- Lying down 1 hour

In a group pen with free choice hay
- Eating 12 hours
- Standing 8 hours
- Socializing 2 hours
- Lying down 2 hours

In a stall with 3 feedings per day
- Standing 18 hours
- Eating 3 hours
- Lying down 3 hours

Horses do not need heated water and generally prefer cool water to warm water. During winters with freezing temperatures, most horses do well drinking cold water but they often need our help breaking and removing ice on creeks, ponds, troughs, or buckets.

Alternatively, freeze-proof buckets and tank heaters can be used as long as they are carefully monitored for excess heat or electrical shorts. I provide free access to clean, fresh, naturally aerated water, such as from a spring or creek, or freshly drawn water in a bucket or tub.

Essential Salt and Minerals

Depending on the season, activity level, and individual metabolism, horses require salt and minerals in order to replenish electrolytes.

Wild horses find natural salt and mineral deposits and sometimes eat soil along with them. For domestic horses, it is best to provide blocks of free-choice plain white salt (sodium chloride), trace mineral salt (red), and possibly calcium and phosphorus supplementation. That way, a horse can select whichever he wants.

Avoid blocks with a high molasses content. Some horses will eat them in a matter of days, thereby ingesting too much salt.

★   ★   ★

Electrolytes *are salt ions such as sodium, chloride, potassium, calcium, magnesium and other minerals that are necessary for various body functions.*

★   ★   ★

Procreation

Wild horses have a strong drive to perpetuate the species: you might argue that it is the reason for their existence. Although domestic horses still have those sexual urges and behaviors, they are usually not free to breed. I will not go into breeding behaviors in this book (see Recommended Reading, page 178), but I will discuss the role of the sexes in herds and the sexual characteristics as they relate to keeping and handling domestic horses.

Socialization and Routines

Horses gather in herds for protection and socialization. There is perceived and real safety in numbers. When possible, keep horses in herds or in bands on pasture. If this is not possible, design facilities so horses are near other horses or can see or hear other horses. Other animals or people can also provide companionship.

Horses are most content when they are allowed to perform their daily routines. Since they evolved as wanderers, roaming as they ate and drank and looked for shelter and safety, horses crave movement and need to mosey around and get regular low-level exercise. When a horse is confined, unable to exercise or flee from danger, he is not content and can panic. Domestic horses require daily exercise and living quarters that are safe and non-threatening.

Since horses are creatures of habit, they prefer to eat, drink, rest, and perform other regular activities at particular times. This ensures digestive health and mental contentment. (See Routines, chapter 5, for more information.)

★ ★ ★

Socialization *is the development of an individual and his behavior through interaction with others of the same species. Horses naturally live in small groups called* **bands**. *In the wild, a breeding band of mares is called a* **harem**, *while a* **bachelor's band** *consists of all male horses.*

★ ★ ★

What Horses Don't Like

Horses have a long list of things that they don't like. Most of them disrupt their eating or safety.

★ Horses do not like to be afraid. If a horse feels threatened or cannot resolve a confusing situation, he will grow fearful and will be likely to panic.

★ Horses do not like physical pain, yet it is surprising how they often tolerate it. When a bit is yanked or a saddle doesn't fit, it is no wonder that a horse might try to rid himself of it; yet many endure pain inflicted by inexperienced or ignorant humans.

★ Horses do not like inconsistency. They are more content knowing what is expected of them and knowing that if they behave a certain way, the reaction from the human will be the same each time. It is confusing and frustrating for a horse to be treated one way today and a different way tomorrow for the same behavior.

★ Horses don't like surprises, although they can learn to become more tolerant of them. Loud noises, such as backfire, gunshot, or dynamite; mysterious sounds, such as rustling plastic; and sudden movements, such as an umbrella being opened alongside the horse, elicit the startle response, which can turn into a full-fledged spook.

★ Horses don't like restraint and restriction, because these take away their ability to flee, but they can learn to tolerate and not fear this. When you tie or cinch a horse or confine him in a stall or trailer, you are restraining him.

★ Horses don't like isolation. Since they are naturally gregarious, they don't enjoy being alone, although, once again, they can adapt to it.

★ Horses do not like being chased, because they are prey animals, and dogs, large cats, and humans are all predators. That's why if you are trying to catch a horse and he turns and starts walking away from you and you keep walking after him, you are confirming, in his mind, that you are a predator stalking its prey.

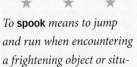

*To **spook** means to jump and run when encountering a frightening object or situation.*

*To **spook in place** means to show fear without moving the feet. This is also called the **startle response**.*

*To **restrain** is to prevent a horse from acting or advancing by using psychological, mechanical, or chemical means.*

Humans and Horses

In general, humans tend to want to dominate things physically. It makes more sense, however, to use our minds, our most powerful tools, to help us become better leaders. Horses naturally follow good leaders. They can also be good mirrors of their humans. A horse's behavior and actions often indicate whether the person handling him is passive, assertive, or aggressive.

No fear. Blue is relaxed and confident as she crosses the crackly orange plastic.

A **passive** horse trainer lets a horse choose what is happening. In the very first interaction between a person and a new, untrained horse, this has some merit, because it helps the horse feel less threatened. If the trainer remains passive as things progress, however, the horse will not respect the trainer and may come to distrust her. A passive person is not well defined, and horses like to know where things stand.

An **assertive** horse trainer is straightforward and confident and lets a horse know where she stands. If an assertive trainer is fair, consistent, and reassuring, a horse quickly learns to respect and trust her.

An **aggressive** horse trainer often wants to win at all costs. Since she tends to feel superior in general, an aggressive trainer views her rights and needs as more important than those of the horse. Because the end goal is often the highest priority with an aggressive trainer, she sometimes hurries and uses force instead of tact.

Human Characteristics

If you understand human behavior, you'll have a better chance of getting along with horses. Some human characteristics dovetail well with horse behaviors, and others are diametrically opposed. Where do you fit?

Horses are very large and potentially dangerous animals, and people react to them in various ways. Since it's human nature to dominate other people and animals, an aggressive attitude can surface when we work with horses. Some people who are not normally aggressive become so because they are actually afraid of horses and shift into a defensive mode. Men who are new to horses often seem to feel that they must prove they can control the beast, especially if someone else is watching. A contest of force develops: "I am going to load this horse into this trailer, no matter what!" Ego, male hormones, and fear — a bad mix.

Sadly, there are also some horse owners who don't respect their animals and take advantage of their generous natures, treating them unfairly or cruelly. Since those people are unlikely to be reading this book, I won't spend time discussing that type of relationship except to caution you: You will see a lot of horse handling and training that is pretty crude and insensitive, but it might be disguised with euphemisms that make it sound as though it is good horsemanship. Beware!

To avoid dangerous conflicts, recognize that you have an ego and determine whether it is of a healthy size. If it tends to puff up, figure out a way to park it somewhere before you head to the barn, or better yet, give it a tune-up or major overhaul. Pride in a job well done is appropriate, however, especially when it concerns horse care and training. That is the sign of a healthy ego.

Modern humans are on a time schedule — we want results and we want them now! We want to know what we have to buy or do to make our horse perfect. In reality, it just doesn't happen like that. The more you give yourself to the horse, the more you will receive in return.

Sassy had a progressive training program and I never asked her to do anything dangerous. That's why she put her trust in me and overcame her inherent fears, such as stepping on this shiny silver tarp.

Dealing with Fear

To know horses is to realize they are not to be feared and that the ideal level of domination is fair and respected leadership. People who fear horses either avoid them altogether or, when they are around them, are so timid that they don't take any action. I've seen middle-aged women in the middle of riding lessons become so petrified that they might do something that could hurt the horse, hurt themselves, or cause them to be criticized by their instructor that they just freeze into total inaction. Such passivity may seem safer than an aggressive attitude, but it is not effective when it comes to working with horses. A continuous series of actions and reactions is required for the

★　　　★　　　★

*The **near side** of a horse is his left side. The **off side** is his right side.*

★　　　★　　　★

wonderful horse-human dance to take place. The trick here is to start with simple things and do them well, never overfacing yourself or your horse.

If a person is so fearful that when a horse acts, she is permissive or submissive, the situation can get dangerous in a hurry. I'm picturing a friend of mine, a tiny woman taking her daily walk through her own pasture where three horses are boarded. Somewhat afraid of horses in the first place, she tried to "make friends" with them. She felt she could please the horses and keep them from "attacking" her by giving them treats out of her pockets. This worked safely twice, but by the third day, the horses became so eager for the treats that they started crowding and jockeying for position close to the woman. Now my friend had a reason to be fearful, because she had created a dangerous situation.

Another common human response to horses is to treat them as though they are human babies or pet dogs. It's dangerous and unfair to think of a horse in human terms. They simply do not think like we do. If you expect them to react the way your husband or mother or best friend does, you will be disappointed and confused. So will the horse. Similarly, thinking a foal is like a big dog and letting him jump up and roughhouse with you will really backfire once he is ten times bigger. Horses are horses and are most content when treated as such.

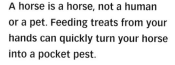

A horse is a horse, not a human or a pet. Feeding treats from your hands can quickly turn your horse into a pocket pest.

How to Become Part Horse

The first and most important prerequisite for becoming part horse is simple, but you either have it or you don't. It is a deep love, respect, and admiration for horses. That is the basis on which all else is built. It also helps if you have an affinity for the particular horse that you are working with. That is an interesting point, though, because when people visit our ranch and see our seven horses, I'm often asked which one is my favorite. I answer that each one is my favorite, and it is absolutely true. Each and every horse, when I am working with him or her, is my best buddy. I see that particular horse's unique qualities and enjoy his or her own way of expression.

I've heard people call their horses "old snide," "blockhead," "dumb-blood," and so on. Even though a horse can't understand the words, there is no mistaking the inflection and intention. What's worse, the horse owner is reinforcing to himself what he thinks about the horse. None of this is appropriate if you are trying to get inside a horse's head.

If you have an inherent regard for horses and a positive attitude, the foundation for a good relationship is laid. The rest is a treat — a lifelong interactive study. Work with as many horses as you can, as frequently as you can, and try a wide variety of activities. In that way, you will come to know horses. As you work, you will develop horse sense, savvy, feel, and timing.

Although reading books, watching videos, and participating in clinics can point you in the right direction, it takes hands-on time and experience to develop feel. Feel is a combination of sensitivity, intuition, and perception that helps you know what to do and when. Some people develop feel quickly; for others it takes a longer time, possibly requiring a major shift in personality to achieve.

Sometimes there is a mental or emotional block that prevents a person from developing feel and timing. I've seen people work so hard that they try to force it to happen rather than allowing it to happen. Once there is a breakthrough, however, the experience can be profound.

As you learn to read horses, look for subtle positive signals, such as small ear movements, straightening of the body, lowering of the head and neck, slight weight shift, leaning, reaching, difference in breathing, tension in the lips and mouth,

To become one with your horse, you need to develop feel. Feel comes from respect, openness, and practice, which develops timing and balance.

and licking. Pay attention and you will soon know what these things mean. Body language is described in chapters 4 and 5.

When you learn to respond the way a horse does, you'll develop a system of communication that you can use to encourage or discourage him. You might encourage or invite a horse to continue his behavior with verbal praise ("good boy") or body language, such as backing off pressure (yielding) or stepping back. You might discourage a horse's behavior with a step toward him, a weight shift, or a gesture.

Horses have so much to offer that it is well worth the investment of time and self to learn all about them.

The Benefit of the Doubt

If you know a horse, and he is a good horse, but all of a sudden he does something out of character, there probably is a concrete explanation for his behavior. Give him the benefit of the doubt. He most likely either senses something you can't sense or he is experiencing physical or mental stimuli that don't affect you.

When one of my horses behaves oddly, I attribute it to horse behavior, not to bad behavior. I look at the environment, the weather, and my horse to try and figure out what is happening. A graphic example is when Aria turns her head, intently looking in another direction than where we are headed. It is often a challenge to see or hear what she is reacting to. On the occasions where I can hear a faint sound or see some movement, I feel as though I am closer to becoming the horse.

The Horse's Senses

HORSES HAVE CERTAIN PHYSICAL FEATURES that cause them to think and act the way they do. In addition, they have ingrained behavior patterns that tell them what to do and when. Learning about these can help us understand the nature and spirit of the horse.

An alert horse carries his head up, with his ears forward, his face line about 45 degrees in front of the vertical, and his nostrils open and actively taking in the scents in the air. The free, alert horse changes the position of his head to view distant and near objects.

Vision

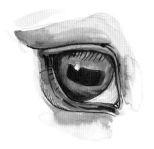

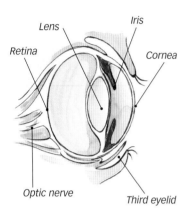

Lens   Iris
Retina   Cornea
Optic nerve   Third eyelid

Although the structure of the horse's eye is similar to that of many mammals, it is almost the largest eye on any living creature, including whales and elephants. A horse's eye is twice the size of a human's.

A horse's vision is quite different from ours. We have a 180-degree field of vision and because our eyes are on the front of our heads, we use binocular vision. The position of a horse's eyes on the sides of his head affords him greater peripheral vision, similar to that of other prey animals. Horses see with both monocular and binocular vision. Monocular vision means that each eye has a separate field of view. With each eye, the horse can see to the front, to the side, and to the rear. Depending on the size and placement of his eyes, a horse's field of view on each side ranges from 130 to 140 degrees, for a total of 260 to 280 degrees of monocular vision.

Binocular vision means that each eye supplies an image and they are superimposed to create a single three-dimensional picture. In order to use binocular vision effectively, a horse must be able to move his head and neck freely. In the horse, the binocular field is 75 to 95 degrees directly in front of his face. Coupled with the monocular fields of vision, in most situations, this gives the horse a 345- to 355-degree field of vision, leaving about 5 to 15 degrees of blind spots.

Because of the placement of his eyes on the sides of his head, a grazing horse can see almost a 360-degree view.

When a horse grazes, constantly moving his head from side to side, he virtually has a 360-degree field of vision. His view to the rear is blocked only by his relatively narrow lower legs, so he needs to slightly turn or rotate his head as he grazes to see behind himself.

Primarily because of monocular vision, we must show everything to both sides of the horse. Don't assume that if he accepts something on the near side, he will automatically accept it on the off side. Routinely work your horse from both sides.

Horses tend to be more farsighted than humans. Try observing a horse as he approaches an object that is initially far off in his binocular field of view. As he draws closer, just a few feet before he begins to pass the object, he might suddenly spook, veer, or want to turn and face it. This is the point when the object is leaving the binocular field and entering one of the monocular fields, and it can cause the horse concern. Whether at that point the image is neither here nor there, is fuzzy, or jumps, we can only speculate. Judging by the horse's behavior, though, it seems that there has been a rough visual transition.

When you ride a horse toward an object on the ground, depending on the length of rein you offer, he can keep a close eye on the object until it is about four feet in front of him. At

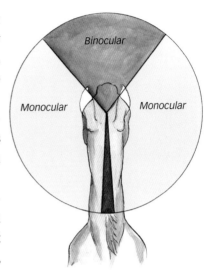

Horses, like other prey animals, can see with both monocular and binocular vision.

From afar, this horse looks at the slicker with both eyes (binocular vision). As he passes the object, he will view it with his right eye only (monocular vision). This is when a horse is most likely to spook — when the object passes from one field of vision to the other. If the rider allows his horse a free rein to get a good look, the result is no spook!

that point, it likely starts getting fuzzy and then enters one of his blind spots. So just as he is about to walk over or through it, he may try to change his head position to get a better look. He might try lowering his head and flexing at the poll or tilting his head to one side or the other. Also, he might try to identify the object by touch or smell. None of this is possible if the horse is being restricted with a short, tight rein. That's why a severely restricted horse can become frightened and confused.

Although a horse has a heightened awareness of motion in the periphery of his vision, he can't focus on the details there; his visual acuity is poor on the sides. A sudden or odd movement can thus cause him to spin and face the object. That's why it is necessary to accustom a horse to objects moving in and out of the different zones, from in front of to the side of the head, and vice versa. A good sacking-out program (habituation) accomplishes this. (See Chapter 9.)

To habituate Seeker to objects moving in the periphery of her vision, Richard flaps a paper feed sack in front of, to the side, and behind her.

Horse "Movies"

Part of the reason that a horse is afraid of things behind him and on his back are the "movies" that continuously play in his subconscious. They go something like this: "As the horse is walking peacefully along, a predator darts out of nowhere and chases him, biting his legs or flank to bring him down." Or Feature B: "A mountain lion waits till he is in the horse's blind spot, then jumps from a rock or tree above and grabs onto his withers." It is no wonder that we have to convince a horse that it is safe to let us handle his legs or to let us sit in his blind spot and ride.

A horse's blind spots include the areas directly behind, below, above, and in front of him.

Blind Spots

As mentioned, horses have very good peripheral vision, especially with their heads down. When a horse's head is up, however, he has several significant blinds spots:

- ⭐ Directly behind him, in the area of his tail
- ⭐ Directly under his head or nose
- ⭐ On his back, in the vicinity of the withers
- ⭐ Directly in front of his forehead

That's why if you come up behind a horse, he will either reposition himself or turn his head to one side so he can see you. If you or a dog or another horse surprise him and he can't move, he might kick at what he senses is there but can't see.

The blind spot directly in front of a horse's forehead is why if you reach suddenly to pet his face, he may spook and become head-shy. The blind spot under his muzzle explains why a horse might accidentally eat your fingers instead of the treat that you are offering. He can smell the treat but can't see it.

A horse's legs are his means of escape and preservation, so that's why he is skeptical about stepping in, on, or over things unless he is allowed to inspect them first. The more experience a horse has, the more trusting he becomes about stepping into something in his blind spot.

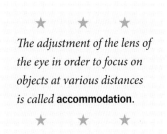

The adjustment of the lens of the eye in order to focus on objects at various distances is called **accommodation**.

Visual Acuity

The ability to focus and see sharpness of detail and contrast, known as visual acuity, is determined in a large part by the number of photoreceptors in the retina. Photoreceptors are specialized cells on the retina that are sensitive to light and make vision possible. Photoreceptors in mammals are rods and cones. Rods are sensitive to changes of light and dark and movement. Cones are sensitive to color but only work in bright light.

Horses have a narrow band (called a visual streak) along the horizontal axis of the eye that contains a high density of photoreceptors, in contrast to the small round spot in a human's eye. This means that when a horse is able to move his head and neck without restriction to focus an object on the visual streak, he has the potential for great visual acuity and especially motion detection. Above and below the visual streak, however, the ability to focus is not as good. So if a horse's head is restricted, his vision is hampered.

Other factors that affect acuity are the shape of the eyeball, the power of accommodation, the elasticity of the lens, and the strength of the ciliary muscles. The horse is thought to have a less elastic lens than a human does, and the lens gets cloudy and less elastic with age.

Taking all of these things into consideration, horses do a pretty good job of focusing on far objects when unrestricted, but focusing on near objects is somewhat problematic. This makes sense in an evolutionary way, because the horses that were successful at recognizing danger from afar and fled are the horses that survived and perpetuated the species.

Light and Dark Adaptation

A horse's rate of adaptation is slower than a human's, meaning that it takes longer for him to adjust from light to dark and vice versa. This is because the shape of a horse's pupil changes from narrow and horizontal (in bright light) to a larger oval or rounded rectangle (in low light). When you lead a horse from a dark barn into the bright sunlight or from the bright sunlight into a dark horse trailer, he might stop at the entrance, asking

The visual streak is a band along the retina that is packed with photoreceptors. This means that the horse can see very clearly in a horizontal band across his field of view, but less clearly above and below the visual streak.

★   ★   ★

When discussing vision, **acuity** *means keenness or sharpness. The* **power of adaptation** *is the ability of the eye to change with varying light intensities.*

★   ★   ★

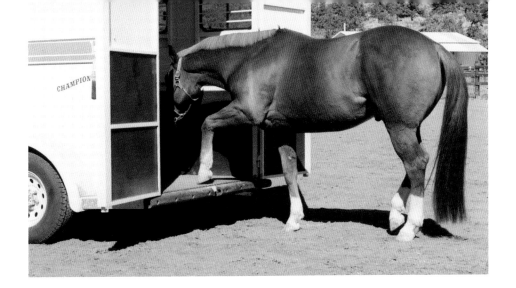

for a few more seconds to adjust to the light change. Your eyes change more quickly, so you are ready to rush in, but if you can spare a few seconds, your horse will feel more secure.

Although horses take longer to adapt to changing light conditions, their range of adaptation is better than humans. This is due to the sheer size of a horse's eye and the large retina surface for light reception. Also, a horse's pupil can dilate six times larger than a human's. Horses have night vision as good as an owl's or a dog's but not as good as a cat's or a bat's. A horse's enhanced night vision is due to the tapetum lucidum, a mirrorlike, fibro-elastic layer on the lower half of the interior of the eye. It reflects and essentially doubles the amount of light sent to the retina. The tapetum lucidum's metallic luster is what causes shine when a light is flashed into a horse's eyes in the dark. So next time you ride your horse on a dark night, know that he can see better than you can because of the special features in his eyes.

And if you've ever "been through the desert on a horse with no name," your steed probably tolerated the bright light better than you did (unless you were wearing polarized sunglasses). That's because horses have built-in sunglasses. Corpora nigra (also called iris bodies or brown bodies) are cloud-shaped, pigmented structures that hang over the iris, partially occluding light from entering the pupil directly. The pupil's narrow, horizontal configuration in bright light tends to decrease the amount of light entering the eye from above (the sun and sky) and below (reflection of the sun on the ground or sand). In addition, a horse's long, downward-angling eyelashes help act as a sunscreen or visor.

It takes longer for a horse's eyes to adjust from bright light to darkness than it does for a human's.

Depth Perception

Since depth perception is possible only in the binocular field, and since a horse's binocular field is much narrower than a human's, a horse's depth perception is not as good as ours. When a horse is free to raise his head to really look at something, he is able to use the area of his eye with the greatest depth perception. Horses are naturally poor at judging distances visually but can be trained to "know" distances, such as for jumping, by using specific training techniques.

Color Vision

The horse's eye has two types of cone cells in the retina that are sensitive to color whereas a human has three types of cone cells. Whether horses can see colors as we do is still widely debated. Most researchers agree that horses do have more than "shades of gray" discrimination but not the color range of humans. Exactly which colors horses see is still unresolved.

The Third Eyelid

A horse has an upper eyelid and a lower eyelid that protect the eye, but in addition, he has a third eyelid (called a nictitating membrane). The third eyelid is located in the inner corner of each eye between the eyeball and the lower lid. When irritants are present, the third eyelid moves quickly across the eye to protect and wipe off the cornea. It also contains a gland that secretes tears to wash and lubricate the eye.

The Eyes Have It

Most breed standards state that large, dark, wide-set, prominent eyes, placed well on the outside of the head, are desirable. Conversely, light-colored or small, recessed "pig" eyes, close together and located on the front of the head, are undesirable. Dark-colored eyes are thought to be less sensitive to bright light. Wide-set eyes are usually coupled with a wide forehead, indicating more cranial capacity and a better temperament. Prominent eyes contribute to a greater field of vision.

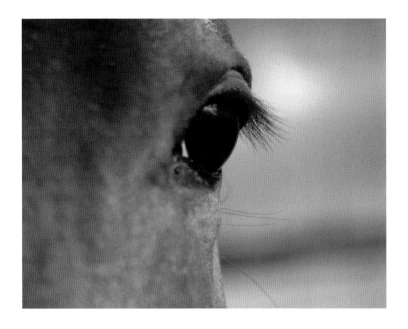

The horse's eye is beautiful and functional. The thick, downward-angling lashes offer protection from debris and act as a sun visor.

Horse Tears

Tears from the lachrymal gland and the gland of the third eyelid wash over the eye, collect in the lower lid, and then flow through a canal (the nasolachrymal duct) that empties through an opening in the floor of the nostrils. If the ducts become plugged, tears run down the horse's face rather than into the nostrils. If age, injury, infection, or dust clogs these ducts, your veterinarian may be able to open them and restore their function.

MOUTH AND NASAL PASSAGES

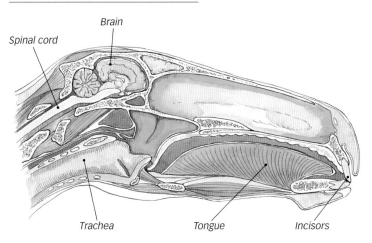

Brain

Spinal cord

Trachea   Tongue   Incisors

Nasolachrymal duct

Hearing

When a horse hears something in the distance, he might stop eating and stand perfectly still, head up and ears forward, completely immobile, as if receiving a transmission from planet Equus. Then, just as suddenly, he might return to eating as if nothing had happened; or he might shift into an excited flight mode, increasing his distance from the object he heard.

A horse's hearing structures are similar to those of other mammals, except that horses' ears are the most mobile of any domestic animal. These large, movable funnels are able to twist nearly 180 degrees from front to back as they focus on and gather sounds. It is generally thought that a horse's hearing is better than a human's in several ways.

Pitch

Horses can hear higher frequencies (pitches) than we can. They can hear low frequencies with their ears, and they can sense even lower frequencies through their hooves and their teeth when grazing.

Frequency is measured in hertz. One hertz is equal to one vibration per second. A sound like the human heartbeat has a low frequency or pitch. A special whistle used to call a dog has a high frequency or pitch, meaning that it has a fast vibration.

Humans can't hear all frequencies: the range for a healthy young person is 20 Hz (foghorn) to 20,000 Hz or 20 kHz (boatswain's whistle). We are most sensitive to sounds in the 1,000 to 4,000 Hz range. The human voice generally has a pitch in the 500 to 2,000 Hz range, although male vowels can be lower than 500 Hz. In general, vowels are below 1,000 Hz and consonants are in the 2,000 to 4,000 Hz range.

With age, we lose our ability to hear high frequencies. By middle age, the highest we can hear is about 12 to 14 kHz. Men lose their high-pitch hearing sooner than women do.

Volume

Horses can hear sounds from greater distances than we can, even several miles away, depending on the wind. And it is generally thought that they can hear (and feel) lower-volume tones than humans and that they are more sensitive to loud noises than humans.

★ ★ ★

Frequency *is a more technical term for pitch, or the number of vibrations a sound produces per second. One kilohertz (kHz) equals 1,000 hertz (Hz).*

Infrasound *refers to any sound with a frequency below a human's audible range of hearing (i.e., less than 20 Hz).*

Ultrasound *is any sound with a frequency above a human's audible range of hearing (i.e., more than 20 kHz).*

★ ★ ★

A horse has the most mobile ears of any domestic animal.

How Horses and Humans Hear Pitch

All ranges listed here are approximate, because various methods of research produce various results. In addition, with animals it is hard to determine when something has been heard.

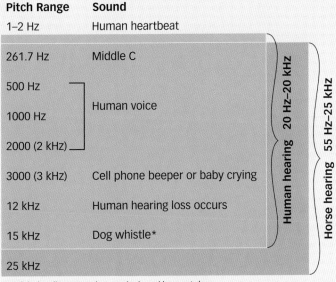

Pitch Range	Sound
1–2 Hz	Human heartbeat
261.7 Hz	Middle C
500 Hz	Human voice
1000 Hz	
2000 (2 kHz)	
3000 (3 kHz)	Cell phone beeper or baby crying
12 kHz	Human hearing loss occurs
15 kHz	Dog whistle*
25 kHz	

Human hearing 20 Hz–20 kHz

Horse hearing 55 Hz–25 kHz

* might be silent to us, but can be heard by most dogs, cats, horses, dolphins, and whales

Who Hears What?

★ Range of hearing for humans in their prime (as young adults) is about 20 Hz–20 kHz with maximum sensitivity in the 1–4 kHz range.

★ Range for horses in their prime (five to nine years of age) is 55 Hz–25 kHz or more, with maximum hearing sensitivity in the 1–16 kHz range.

★ Range for dogs in their prime is 40 Hz–60 kHz.

★ Range for cats in their prime is 45 Hz–85 kHz.

★ Bats can detect ultrasound frequencies as high as 120 kHz; dolphins, 200 kHz.

★ Elephants hold the infrasound record, with a range from 5 Hz–10 kHz.

Volume is measured in decibels (dB), which is a logarithmic unit that represents the energy of the sound. Decibels aren't like ordinary numbers. For example, 20 dB has ten times the energy of 10 dB, and 30 dB has one hundred times the energy of 10 dB.

Horses have survived because they developed keen hearing, so it is no wonder that they are always listening, are innately suspicious, and can be easily startled by various sounds.

Loud noises with high decibel ratings, such as gunfire, vehicle backfire, and diesel truck engine brakes, can cause any horse to startle and possibly spook. This fear can be overcome with systematic conditioning, however, as evidenced by the successful use of horses in the military and police forces.

Quiet barn noises of the type that take place during grooming and tacking up fall into a pleasant range somewhere around 20 to 35 decibels. Noises of 85 decibels or greater, however, such as loud music or a tractor or truck pulling up to the barn, can be unsettling and even harmful to hearing.

Sounds That Worry Horses

When traveling, horses can become very anxious from constant truck, trailer, and road noise — the rattling of stall dividers and doors, the sound of the engine, and other traffic noise. The more a horse travels, the more he can become accustomed to the noise, but in the early stages, you can dampen the effects by putting cotton in his ears.

When a horse shows an aversion to having his bridle path or throatlatch clipped, he could be reacting to the vibration of the clippers, but more often it is the buzzing sound that bothers

Volume of Common Sounds

Here are some everyday sounds that you and your horse hear. The red line indicates the level beyond which pain, damage, or death can occur.

Sound	Volume in Decibels (estimate)	Sound	Volume in Decibels (estimate)
Hearing threshold (human)	0	Tractor at 80% throttle	100
Human breathing	10	Lawnmower, firecrackers	100
Rustle of leaves	20	Thunder	100–130
Whispering	20–30	Motorcycle, chain saw	110
Mosquito flying	40	*OSHA requires protection*	115
Normal talking	40–60	**Tickling sensation**	120
Singing birds	60	Rock concert, snowmobile	120
Busy traffic	65–80	**Unsafe; pain occurs**	130
Vacuum nearby	70–80	Jackhammer or gun	130
Train nearby	65–90	Dynamite blast	140
Telephone dial tone	80	Jet nearby	150
With repeated exposure, hearing damage begins	85	**Physical damage occurs**	150
Heavy traffic	90	Shotgun blast, muzzle end	160
OSHA regulations begin	90	**Rupture of eardrum**	190
		Death	200

him, resembling a horde of insects attacking one of his blind spots. Again, the instinct to survive tells the horse to avoid clippers. It is easy to teach a horse that clippers are no threat by using systematic desensitization. Unfortunately, it is also easy to teach him to fear the clippers if you physically restrain him and give him a sharp whack when he raises his head.

Wind, in and of itself, is noisy. It also brings with it more sounds for the horse to process, and it masks other sounds. A horse might normally hear sounds from one-quarter mile away. With a 15-mile-per-hour wind, he might hear sounds from one-half mile or more upwind and very little from downwind. It is no wonder horses are uneasy in the wind.

Sounds That Trigger Anticipation

Horses link certain sounds to specific activities, such as feeding. Just like Pavlov's dogs that salivated when they heard a bell ring before being fed, horses anticipate being fed when they hear sounds they know precede feeding. Here on our ranch, the horses respond by becoming more active physically or calling for feed when they hear:

- ⭐ The very first word or morning yawn from the house

- ⭐ The back door to the house opening and closing

- ⭐ The barn door sliding open or the feed room door opening or closing

Horses also hear subtle precursor sounds that let them know another horse is moving through their territory. Our horses often call out when they hear gates or stall doors opening or closing, which signals the movement of horses and means it's time to say hello or good-bye. When they hear the rattle of stall dividers in a horse trailer way off in the distance, they are alerted and keenly focus on its direction. When it passes by our ranch, if there is a horse in it, they often exchange vocal greetings as the trailer goes by.

When a cattle drive is still several miles away, they will be preoccupied with that direction. Sure enough, an hour or so later, the cows start filing past our lane. In many cases, our horses make better watchdogs than our Rottweilers do, letting

Horses are restless in the wind because of what they hear and what they can't hear.

Sherlock stops and waits, watching and listening for my cue to either go forward or turn.

us know by their alert posture and ears pointed in a particular direction that something unusual is happening in our valley.

Horses can discriminate between various human voices and words. (See more on this in chapter 8.)

Speaking So Your Horse Will Listen

A horse's hearing is so keen that loud voice commands are not only unnecessary, they are also rude and counterproductive. Good horsemen can often be observed communicating with a horse in a type of low-level "breaking patter," a term from bygone cowboy days. It describes a type of low-volume mumbling a cowboy might use around a horse that he is training. The soothing tones calm the horse. Hence the term *horse whisperer*.

From my observation, the music or radio talk shows that play in some horse barns are more for humans than the horses. Some horses adapt to the constant noise, tune it out, and relax in spite of it. Since a barn should be a peaceful haven for a horse, however, I believe it would be more appropriate if there were no radio or just soft, soothing music playing.

Smell and Taste

The senses of smell and taste are more highly developed in horses than in humans, and they are closely connected. **Smell** is processed in a horse's moist nostrils, which have a very large surface area. Odor particulates are carried through the air and deposited on the moist tissues, and the information is sent to the brain for decoding.

Taste is processed by papillae on the tongue, throat, and palate. Liquids or solids that the horse ingests pass over the tongue and are either accepted or rejected. Horses naturally like salty and quickly learn to like sweet, but generally don't like bitter or sour.

If a horse detects the flavor of a bitter pill, or suspects that one is coming his way, he will spit it out or try to avoid taking it. That's why we sometimes need to disguise or mix certain medications with something sweet, such as molasses-coated sweet feed, soaked beet pulp, frosting, or applesauce.

Smell is horses' tool of recognition. Their ritual is to smell without being smelled. Just like with dogs, when two horses meet, each one wants to find out all he can about the other horse, without letting the other horse get too close to him. When a horse is meeting a new horse, person, or object, the smelling might need to be very thorough and could take some time, but with a known associate, the sniffing is often ritualistic and quick. After a brief sniff between friends, mutual grooming often follows. After the deeper scent exchange takes place between strange horses, if there is a challenge or threat they might swish their tails, lift their hind legs, pin back their ears, lower their heads, squeal, and possibly strike or kick.

Mutual grooming is reciprocal nibbling along the neck, withers, and back between two horses, usually bonded buddies.

The Flehmen Response

If the smelling ritual includes recognition of hormones in urine, sweat, or other bodily fluids, the odors often elicit the flehmen response. The horse curls his upper lip back, driving the scent into his nostrils and sealing it there, where the odor particles can be processed by the vomeronasal organ (also known as the Jacobson's organ) at the top of the nasal passages. The area is physiologically structured so that there is a direct route between odor processing and the flehmen response behavior. Other things that cause the flehmen response include medicine, blood, perfume, smoke from hot shoeing, cigarettes or hands that have held them. Although to us it is an odd behavior, it is perfectly normal.

In some horses, the flehmen response is much more common than in others. Our big sorrel gelding, Dickens, has an exquisite flehmen response. It is predictable and easily triggered by worming preparations, mare urine, or new feed. It lasts for minutes once it has started. Yet I can never remember

A World of Smells

A horse's behavior is often influenced by odors — subtle smells that we might not detect and strong smells that we might just label foul or pungent. To the horse, they are a storehouse of information. Here are some examples.

Shortly after birth, the mare and her foal participate in an important bonding session through smell and taste. The mare licks the foal, and the foal nurses from the mare. This recognition lasts a lifetime.

Horses mark their territory with manure and urine. Stallions do this most distinctly by establishing and using stud piles (mounds of manure) and by covering a mare's urine or feces with their urine. Any horse, however, when put in a fresh stall, scents it with his or her own body aromas. Perhaps it makes the stall feel more like home. The smell of urine is also the primary way a stallion detects whether a mare is in heat.

Horses are notoriously good at homing, and it has been suggested that their primary means of finding home is by following the trail of scent.

The larger and more open a horse's nostrils are, the more oxygen and information he can take in.

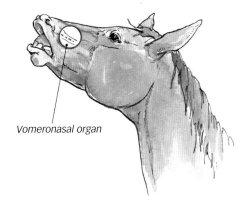

Vomeronasal organ

When processing odors, a horse sends the scent up the nasal passages to the vomeronasal organ.

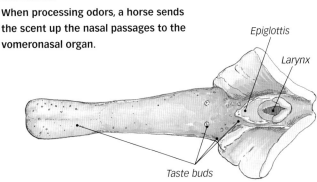

Epiglottis

Larynx

Taste buds

seeing our gelding Sherlock or mare Seeker exhibit the flehmen response. To see if your horse will exhibit this behavior, the next time you clean your gelding's sheath or your mare's udder, let that horse or another smell your glove with the waxy residue on it.

The Scent of Water

The equine sense of smell is so acute that it can make a horse go off water if the source of his drinking water changes. He can detect subtle differences in the mineral content and, depending on the horse, he might refuse to drink unfamiliar water. If offered only unpalatable water, it could be days before he will cave in and drink, and by then he might be severely dehydrated.

If you travel with your horse, you can flavor his water. A week before departure, begin using a small amount of apple

Horses are extremely sensitive to the smell of water, which protects them from drinking from a contaminated source.

juice, flavored gelatin, or another aromatic substance to flavor the water at home. While on the road, repeat the process, so he will be less likely to pick up the differences in water.

Sniffing Danger

Since horses can't regurgitate, it is fortunate their keen sense of smell prevents them from eating harmful substances. The senses of smell and taste protect most horses from the multitude of poisonous plants that populate and border pastures. If there is sufficient high-quality grass forage, a horse will tend to avoid eating odd plants. If there is nothing else to eat, however, a horse will finally give in and eat the usually less palatable plants, some of which can be toxic.

Unless horses are very hungry or thirsty, they generally won't eat feed or water that is "off" — hay or pasture that has been trampled or fouled by other animals or spoiled by rodents, mold, bacteria, dirt, dust, scum, slime, or algae. This is a good thing, because a horse's digestive system is fairly sensitive to many of these agents.

That said, you've probably seen a foal eating Mom's manure. It is thought that this practice enables the foal to populate his intestines with the bacteria necessary for digestion of forage. Disgusting by human standards; normal for horses.

When high-quality pasture is available, horses tend to avoid unpalatable or poisonous plants. Foals and young horses learn what to eat or drink and how to graze by watching their dams or other horses. Once a horse has tried a feed, if it tastes okay and doesn't make him sick, then he has learned to adapt to that feed as nutrition.

What Do Horses Like to Eat?

Various factors can affect what a horse will eat. Wild horses tend to have an inherent wisdom regarding what is good or bad to eat and also what they might need to eat in terms of salt, minerals, or herbs. Our domestic horses may or may not retain that instinct. If a stalled horse that is rarely turned out is suddenly put on a weed lot, he is likely to ingest weeds that could cause colic or worse, just to enjoy eating something green and growing.

Given their druthers, however, horses show some very specific preferences for taste and texture, and what you find out might surprise you. Each horse has a different palate, but I've found that, in general, my horses prefer coarse hay to fine hay (see box below), and they eat more broad-leaved plants, including thistle, on pasture than I'd expect.

Hay Taste Test

When it comes to hay, I'm constantly doing taste tests with my horses. For example, in a recent test, I set out three types of hay:

1 **"Busy hay"** — Very coarse, overly mature orchard grass/brome hay mix that was put up perfectly dry and loosely baled. No seed heads, just coarse pale green stems.

2 **"Green hay"** — Bright green, fine grass/alfalfa mix that was densely packed and put up moist, so it was still a bit damp to the touch but gorgeous to look at. Because of its high moisture content, it was soft and would bend rather than break. It had a unique aroma that I would call slightly acidic.

3 **"Caramel hay"** — Tightly baled, heavy flakes of rich grass/alfalfa hay that had caramelized in the center of the flakes to a deep brown, tobac-colike color. This hay was not moldy or dusty but it was odd hay, with more of the scent of a sweet cigar or burnt Vidalia onion than that of a salad.

By visual inspection, one would expect the horses to prefer the hays in this order: 2, 1, 3, or green, busy, caramel. Almost all of the horses, however, preferred them in this order: 3, 1, 2; or caramel, busy, green. Among our various horses, the caramel hay and the busy hay were often quite close in preference, while the green hay was always left till last, often just lightly picked through and wasted.

So, based on the results, we renamed the hays:

1. **Basic hay**
2. **Acid hay**
3. **Pâté hay,** as in smoked foie gras, an aged hay for connoisseurs.

Touch

When it comes to the sense of touch, just because horses are big does not mean they are dull. In fact, quite the opposite is true: horses are exquisitely sensitive.

Sensitivity differs greatly among individual horses, depending on the thickness of their skin and hair coat and the type of receptors at various parts of the body. While some cold-blooded horses can show duller reaction times and slower response to touch, most saddle horse breeds, which are a mixture of cold- and hot-blooded breeds, are quite sensitive. A horse's skin and underlying muscles react the way yours do to light touch; heavy, steady pressure; pain; heat; and cold.

Sensitive Areas

In addition, a horse has particularly sensitive "feelers" that take the place of hands when inspecting things. The whiskers on a horse's lips and nose and around his eyes are antennae that help him detect where he is putting his head, especially in the dark. Since he can't actually see what is in the bottom of the bucket or water trough, his feelers help him get into tight or deep places without hurting his head. That's why it is best not to clip off these feelers for supposed aesthetics, such as in the

★ ★ ★

Horses with ancestors that trace to heavy war horses and draft breeds are often described as **cold-blooded**. *Their characteristics might include more substance of bone, thick skin, heavy hair coat, shaggy fetlocks, and lower red blood cell and hemoglobin values. Horses with ancestors that trace to Thoroughbreds or Arabians are called* **hot-blooded**. *Their characteristics might include fineness of bone, thin skin, fine hair coat, absence of fetlock hair, and higher red blood cell and hemoglobin values.*

★ ★ ★

ZONES OF SENSITIVITY

Knowing where your horse is most and least sensitive will help you choose the appropriate touch, tools, and aids.

Very sensitive

Medium sensitive

Toughest: can be rubbed

show ring. Whiskers are there for a reason — they are a necessary protective feature.

The muzzle is a highly tactile area containing nerve endings, whiskers, and the sensors for smell and taste. Although a muzzle is soft and begs to be petted, how would you like it if someone came up and put their hand in your face? Respect the horse's muzzle as a sensory center and remember that below it is a blind spot. Unfortunately, I've seen many people become injured or lose a finger when a horse instinctively bit what was under his muzzle, either because he felt threatened or because he was anticipating being fed. Try rubbing your horse's forehead, neck, or withers instead of his muzzle and he'll tell you which he prefers.

Horses use their mouths to lip (inspect), lick (inspect), chew (inspect or destroy), bite (destroy), warn, and defend. Items that are commonly bitten or chewed include ropes, blankets, wood, fences, buckets, and the manes and tails of other horses. In addition, horses use their mouths and teeth to groom themselves (legs and sides) and others (mutual grooming).

A horse also uses his hooves to inspect things. He will try to gain a sense of safety, softness, or depth by pawing footing or flooring. Horses usually can sense unsafe, boggy ground well before stepping into it.

How Horses Like to Be Touched

In general, horses like to be rubbed, not tickled or slapped. They enjoy being rubbed on their forehead, neck, withers, back, croup, and chest. The rare horse would solicit rubbing on sensitive areas such as the flank, girth, belly, nose, ears, or legs. So when you first start working with a horse, handle him in places you know he will naturally enjoy and gradually get him used to being handled in his ticklish areas. Every horse has his favorite spot, so experiment with each of your horses.

Since horses love rubbing on many parts of their bodies, it should come as no surprise that they also love to rub their bodies on fences, buildings, and even other horses, often for long periods of time and with rhythmic swaying. This behavior can destroy a mane or a tail head in a single session and can result in ripped blankets or damaged fences or buildings. Regular grooming and allowing a horse to roll usually prevent destructive rubbing habits from becoming established.

This foal has found a solid limb of the perfect height for him to obsessively rub his mane and neck against. By now his neck is almost bald.

How to Pat a Horse

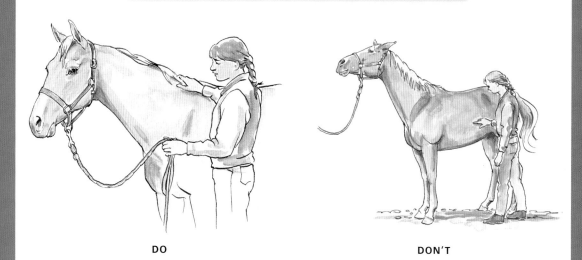

DO   DON'T

When touching your horse, it's better to rub him (left),
rather than slapping or tickling him (right).

DO

DON'T

Horses prefer being rubbed on the forehead (above)
over being dabbed on the nose (right).

Respecting a Horse's Sensitivity

Horses are sensitive to touch, which means we can and should train with very light pressure, not force.

Horses tend to move *away* from light intermittent pressure (it is irritating) and lean *into* heavy, steady pressure (it is comforting). If you've ever tried to move an untrained horse over and resorted to leaning bodily into him, you most likely found that he leaned into you with all of his weight and loved it! When communicating with the horse, whether through a halter or a hand, small taps are much more effective than an all-out tug-of-war. As odd as it might seem, a light tap or a pesky little tickle on the ribs is more effective in making a horse move away than steady pressure. Take a lesson from the flies!

Touch is an active sense, not a passive one. Desensitization results from becoming accustomed to prolonged or repeated stimulus. Here's an example. Where we live, there are very few horseflies, so when a horsefly does land on one of our horses to suck blood, it drives them crazy. One spring, a stallion from Wyoming was visiting our ranch, and he brought along his own entourage of horseflies. They covered portions of his back like shingles on a roof. He didn't react to them at all. He had become desensitized to the stimulus of the flies.

A horse will move away from a light, pesky tickle, whether made by a fly or a whip.

When you ride, you communicate with your horse by contact through his mouth or nose and on his back and sides. He will feel if you are relaxed or tense through your seat on his back and by the feel of your legs on his sides. Although you first need to accustom a horse to being touched all over without becoming afraid (see discussion of habituation in chapter 9), constant stimulation on his body can deaden him to the sensations, leading to the notion of hard-mouthed and dull-sided horses.

If you use minimal intensity in your rein and leg aids, you will develop and maintain a horse's sensitivity. Otherwise, you might habituate him to the aids to a point where he is dull to them and no longer responds. The temptation at that point is to bring out spurs and a more severe bit. There is a fine line between the desensitization that is necessary to make a horse safe and keeping the horse responsive to the application of aids. Therein lies the art of horse training.

Considerate Grooming

Use the appropriate tool in the right place and with the correct technique.
The tools are listed in order from most harsh to most gentle.

Tool	Area of Body	Technique	Comments
Metal curry	Not to be used on the horse at all	Run the mud brush over the metal curry to clean it	Save for using on other brushes, not on horses. The sharp metal teeth would scratch a horse's skin.
Shedding blade	Heavily muscled body parts (neck, shoulder, hindquarters)	Long strokes and light to moderate pressure	Use to remove long, thick winter coat. Sharp teeth can scratch skin and create an open invitation for skin infections.
Sweat scraper	Heavily muscled body parts	Long strokes and light to moderate pressure	Use to remove water from coat after bath or sweat after a hard workout.
Rubber curry	Heavily muscled body parts	Moderate pressure and circular motion to rough up the coat and loosen debris	Choose a pliable rubber curry that conforms to the horse's body and your hand. Avoid hard plastic curries.
Stiff-bristled mud brush	Heavily muscled body parts	Short flicking motion to send dirt and debris off the coat	Either natural or synthetic bristles are okay as long as the ends are not sharp.

Tool	Area of Body	Technique	Comments
Medium brush	Ribs, upper legs	Long strokes to remove final particles and to start laying down the coat	Use also to brush mane and tail.
Very soft rubber curry	Face, lower legs, belly, flank	Light but steady pressure and circular motion	Wherever there is just skin over bone, use a soft rubber curry that has tiny rubber fingers.
Very soft body brush (such as fine horsehair)	Face, lower legs, belly, flank	Smooth strokes and flicks	A soothing finishing brush for sensitive areas.
Cloth, sponge	Face, flank, under tail, and to clean udder and sheath	Wet, let set, then wipe. No scrubbing	Warm water makes cleaning easier and is more comfortable for the horse.
Hand, with or without gloves	Face, ears, jaw, lower legs	Massage or stroke	Use hands to strip water off lower legs. Use grooming gloves with bumps to clean face and area around eyes.

Reflexes

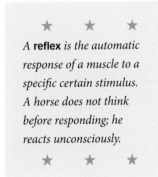
As the horse survived over millions of years by avoiding predators, he developed a set of reflexes that remain with him to this day. Natural selection favored horses that escaped predators. These individuals passed along their highly developed instincts, and today's horses exhibit a vast array of deeply ingrained reflex chains.

Because reflexes are unconscious reactions, they are potentially dangerous. We need to systematically override such reflexes in order to handle and use a horse safely.

Flight is a horse's main means of defense, so a horse's legs are vital to survival. Anything that diminishes his ability to use his legs is a threat. For example, picking up the feet is often a difficult task for novice owners, because a horse has deeply ingrained protective reflexes related to his legs. The goal is first to accustom him to having his leg touched, and then to pick up the leg for a moment and build on that.

The Suck Reflex

When a mare nibbles her newborn foal or the foal rubs his head up against the mare, or a person scratches the foal on the top of the head or over the tail head, it causes the foal to reach with his head and neck, search for the dam's udder, and make suckling motions with his lips. Often, any rubbing along the foal's spinal column or the top half of the head will precipitate oral movement, such as when two horses perform mutual grooming. Because nuzzling and nibbling often lead to biting, when we groom or handle foals and young horses we need to override their natural desire to "mouth" us.

Withdrawal Reflex

The withdrawal reflex is what causes a horse to snap his leg off the ground when it is touched by a predator, a fly, a hand, or clippers. Certainly, you want to preserve your horse's natural protective instincts so he can take care of himself in a pasture, but you also want to work safely around his legs. You need to systematically train your horse to keep his weight on his legs for clipping, grooming, and bandaging. You want to train him to pick up his foot when you ask him to, while still fully expecting him to react quickly if a fly lands on his cannon.

Croup and Perineal Chains

The croup and perineal reflex chains cause a horse (particularly a mare) to clamp the tail, tuck the croup (squat), and possibly kick and buck when the underside of the tail or anus is touched, especially with something cold, such as a neoprene tail wrap or a spray of water. This is an automatic physical response to attack, to unfamiliar tack, or to handling. To this day, my dear 31-year-old mare, Zinger, shudders and squats once every time she is bathed.

These reflexes also make a horse clamp down her tail and tighten her anal sphincter muscle just as you are about to insert a thermometer or palpate. Sassy, my 29-year-old broodmare, has a perineal reflex that has been dubbed "tail of steel." Although she quickly overcomes it and relaxes in response to a soft touch and consideration, she has a very strong and determined instinct to protect her anus and vulva. Sassy has been an excellent broodmare, settling easily with no infections, and foaling well into her mid-twenties. It would seem that her self-protective reflexes have contributed to her breeding soundness and longevity.

Cutaneous Trunci Chain

The *cutaneous trunci* (panniculus chain) is responsible for the rapid, repeated muscle contraction of the skin over the horse's barrel: for example, when a fly lands on his rib cage. This same reflex can make a horse hypersensitive to a rider's leg cues.

Spina Prominens Chain

The *spina* (vertebra) *prominens* chain causes a horse to hollow his back if you run a fingernail down his spine. Improper conditioning or ill-fitting tack, coupled with this reflex, may lead to a hollow back and bucking when the horse is saddled or ridden.

Other Reflexes

Other reflexes commonly seen in horses include ear twitch, eyeblink, tearing, pupil dilation, head shake, saliva production, sneezing, and coughing. If you are aware of the origin and nature of reflex chains, you can design your lessons to help calm your horse's fears and override his reflexes.

> ★ ★ ★
>
> **Aids** *are the means by which a trainer or rider communicates with the horse. Natural aids are the mind, voice, hands, legs, and body (weight, seat, and back); artificial aids include the halter, whip, spurs, and chain.*
>
> ★ ★ ★

Sherlock's strong reflex in response to light pressure on the right side of his topline caused him to bend his neck so dramatically that he could easily touch his hip.

A Map of Equine Reflexes

Reflexes are automatic responses to pressure on or movement of various portions of the body. The intensity of the reaction will vary depending on the horse's physical makeup (thin skin, fine hair coat, hot- or cold-blooded, etc.), temperament, experience, training, physical restriction, degree of relaxation or tension, and how forcefully and with what means the pressure is applied. A willful, resentful, or sullen horse will override his own reflexes and tune out your aids as a means of protection or defense, and nothing you can do will get a reaction.

Normal reflexes are reactions to pressure on various areas. A wild or unhandled domestic horse will show reflex reactions dramatically. A seasoned, trained riding horse may show little or none as the reflexes have been overridden by habituation and training.

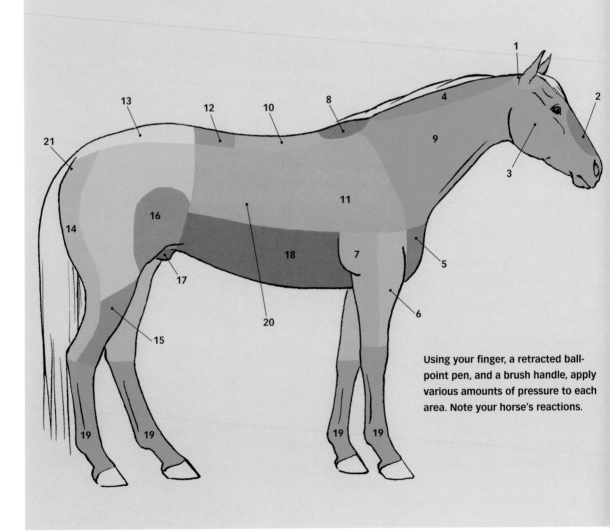

Using your finger, a retracted ball-point pen, and a brush handle, apply various amounts of pressure to each area. Note your horse's reactions.

1. **Poll:** Raises the head and neck; may try to pull away. Can be conditioned to lower head.

2. **Bridge of nose:** Raises head, hollows neck, flips nose up. Can be conditioned to lower head.

3. **Poll flexion:** With no alteration in a normal neck position, an upward rotation of the head via extension at the poll or raising of the head causes forelimb flexion and hind limb extension; a downward rotation of the head or flexion at the poll induces hind-leg flexion and foreleg extension.

4. **Crest:** Lowers neck.

5. **Breast:** Backs up if the head is low. If head is high, reflex is blocked.

6. **Forelimb extensors:** Causes cannon and hoof to move out to the front.

7. **Forelimb flexors:** Causes leg to bend at the knee.

8. **Withers:** Light pressure: lowers head, reaches with head, nibbles if scratched. Heavy pressure: moves away from pain, makes threatening gestures with head and neck.

9. **Tonic neck reflex:** Contracts on the touched side. With no alteration in normal head position: Laterally concave neck induces hind-leg flexion and foreleg extension on side touched. Laterally convex neck induces foreleg flexion and hind-leg extension.

10. **Back (spina prominens chain):** Light pressure on spine from withers to loin will cause hollowing. On left side of spine, will cause spine to curve away from pressure, left hind leg to move forward, and possibly head and neck will curl to the left also. The reverse will occur with pressure on the right side.

11. **Ribs:** Head turns toward pressure, ribs flex away from pressure, nearest hind leg flexes, opposite hind extends, causing sway or crossing.

12. **Loin:** Flattens or rounds back.

13. **Croup:** Tucks tail and hindquarters and rounds back.

14. **Semitendinosus (hamstrings):** Raises leg or kicks backwards.

15. **Gaskin:** Flexes hock.

16. **Flank:** Reaches hind leg forward or "cow-kicks."

17. **Sheath:** Reaches both hind legs forward. Drops croup.

18. **Abdominal muscles:** Contracts belly, rounds back, arches neck, drops croup.

19. **Distal limbs:** Withdraws legs by flexion.

20. **Cutaneous trunci or panniculus (sheet of muscle under skin of barrel) reflex:** Light stroking causes rapid, repeated twitching. Firm, steady pressure causes isometric contraction.

21. **Perineal:** Contact with anus causes contraction of anal sphincter and tail clamping.

Proprioceptive Sense

Receptors located in muscles and tendons constantly send messages to the brain that help coordinate a horse's movements. Horses generally have very good proprioceptive sense, some much better than others.

Coordination is closely coupled to the proprioceptive sense. A horse is well coordinated if his body parts function harmoniously when he performs complex movements such as canter, stop, and turn.

Proprioceptive sense is knowing where parts of the body (such as the limbs) are in relation to other parts and to objects, without being able to see them. Zipper has good proprioceptive sense as he crosses the brightly colored bridge.

Senses at a Glance

VISION
★ Horses have a broader range of peripheral vision and a wider range of adaptation to light than humans do.
★ In a certain zone at far distances, horses have better visual acuity and motion detection.
★ Humans take home the all-around better visual acuity trophy.
★ In general, humans have a faster rate of light adaptation than horses do.

HEARING
★ Horses have more mobile ears than humans do and can pinpoint a sound's location without moving their head.
★ They can hear higher and lower pitches and seem to have greater sensitivity to volume at both ends of the scale.

★ Although a horse's ears are used primarily to take in communication, they are also a means of communication, as will be discussed in chapter 8.

SMELL AND TASTE
★ Smell and taste are more highly developed in horses than in humans.
★ These senses gather information about the environment and other horses, protect the horse, and guide his behavior.

TOUCH
★ Because the horse is a sensitive animal, always start with the least intense signal possible and go from there. Not only does that leave you room to amp up your aids, but it also ensures that your horse is being treated fairly and humanely.

The Physical Horse

IN ADDITION TO THE SENSES, some unique physiology sets the horse apart from humans and other mammals. Understanding these characteristics will help you provide more appropriate handling and better care.

"Strong as a horse" is somewhat of a paradox. While many physical features make the wild horse tough and durable, capable of surviving millions of years, the domestic horse can be very sensitive and vulnerable. Feeding, veterinary care, hoof care, and training practices can make or break a horse. Knowing about the horse's teeth, intestines, back, skeleton, and hooves will help you become a better horsekeeper, trainer, and rider.

Seasonal Changes

Modern man has all but lost his natural connections to seasonal rhythms. Not so the horse. The horse's inherent calendar and biological alarms and triggers automatically guide him through such events as procreation and wardrobe changes.

Breeding Season

Horses are seasonally polyestrous, which means they have a specific breeding season each year, with multiple breeding periods triggered by day length. In the Northern Hemisphere, the most viable breeding months are from spring (April) to fall (September).

A mare's estrous cycle lasts an average of twenty-one to twenty-three days, and during that time the mare is in standing heat (receptive to the stallion) for an average of five to seven days. Gestation length is eleven months. In the wild, foals are generally born in the summer, to take advantage of warm weather and good pasture. All wild and pastured horses tend to gain weight in the summer and fall in preparation for winter, when feed is scarce and many plants are dormant.

Seasonal Coat

In spring, in response to lengthening days, a horse sheds his winter coat and grows a shorter summer coat. In the fall, in response to shortening days, he sheds his summer coat and grows a long winter coat. His hair grows thicker and longer in the winter to protect his body from cold, wet weather, and he gets particularly shaggy under the jaw, along the belly and underline, and on the legs.

A horse's skin secretes a waxy exudate called sebum, a natural waterproofing. Even if rain or snow penetrates the long hair coat, if the sebum is undisturbed it will repel the moisture. Although grooming in the summer distributes natural oils and is good, bathing or deep and thorough grooming in the winter strips the skin of its built-in protection and so is not advised unless the horse will be blanketed. Horses do not require blankets unless they are ill, thin, old, or have no shelter.

Horses have guard hairs that act as antennae around their eyes and on their muzzles. The protective hair inside their ears wards off insects. These hairs should not be clipped off.

Zinger's thick winter coat and long guard hairs keep her skin dry and her body warm.

Digestive System

Horses thrive on grass and water because their digestive system has evolved over millions of years as a nomadic grazer. Knowing some of the features of the equine digestive system will help you become a better manager.

Teeth

At five years of age, a horse has a full mouth of adult teeth with substantial reserve crowns below the gum line. The teeth continue to emerge until the horse is twenty years of age. As he chews, from side to side as well as up and down, his teeth wear in a unique manner that requires regular dental care. (See more about teeth in chapter 7.)

Intestines

The cecum, part of the large intestine located on the right rear side of the horse, acts as a fermentation vat, breaking down cellulose with the help of microorganisms. The cecum can become impacted and cause colic. The pelvic flexure is a rather sharp turn in the large intestine on the left rear side of the horse. It can also become blocked and cause colic. A horse with colic might turn and look at his flank, bite his sides, paw, lie down and get up, thrash and roll while down, sweat, pace, and won't eat or drink.

Spleen

The spleen acts as a red blood cell (RBC) reservoir supplying extra RBCs, which carry oxygen, during exertion. Release of adrenaline triggers the release of the extra red blood cells.

No Return

Because of tight sphincter muscles in the esophagus, horses cannot regurgitate, so what goes in must continue through the digestive tract or result in rupture or impaction. When a horse is blocked at both ends and is suffering from gas, impaction, or the production of toxins, it can lead to colic, one of the leading causes of death in the horse.

THE HORSE'S DIGESTIVE SYSTEM

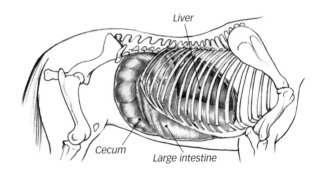

Liver

Cecum   Large intestine

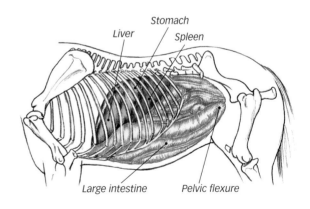

Stomach

Liver   Spleen

Large intestine   Pelvic flexure

Skeletal System

A horse's skeleton has approximately 205 bones. The reasons for approximation include:

- ✦ Fusion of some bones, such as the five sacral vertebrae, as a horse ages

- ✦ Breed variation: Arabians might have 38 ribs but one fewer lumbar vertebra

- ✦ Different number of tailbone vertebrae, ranging from 15 to 21

Weight-Carrying Ability

A horse's body isn't really designed to carry extra weight, but it can by virtue of its suspension-bridge features. How much weight can a horse carry? This will depend on several factors including the horse's weight, bone, conformation, breed, condition, type of riding, rider's skill, and type of saddle used.

An often-quoted rule of thumb is that a horse can carry 20 percent of his weight. This would mean a 1,200-pound horse

EQUINE SKELETON

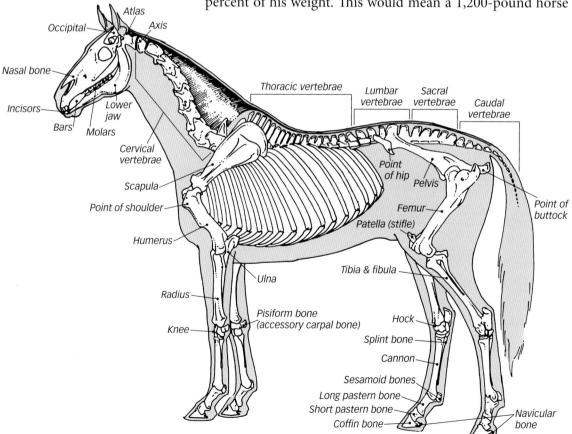

Atlas
Occipital
Axis
Nasal bone
Incisors
Lower jaw
Bars
Molars
Cervical vertebrae
Scapula
Point of shoulder
Humerus
Radius
Knee
Pisiform bone (accessory carpal bone)
Ulna
Thoracic vertebrae
Lumbar vertebrae
Sacral vertebrae
Caudal vertebrae
Point of hip
Pelvis
Femur
Patella (stifle)
Point of buttock
Tibia & fibula
Hock
Splint bone
Cannon
Sesamoid bones
Long pastern bone
Short pastern bone
Coffin bone
Navicular bone

could carry 240 pounds, which would include the rider plus tack. Horses with denser, larger bone might be able to carry more than the 20 percent. Bone is determined by measuring the circumference of the foreleg just below the knee. Average is about 8½ inches for a 1,200-pound riding horse. If a horse has lighter bone, he would likely be able to carry less than 20 percent. If he has heavier bone, he would likely be able to carry more than 20 percent.

Equine Bones: A Breakdown

Skull	34	
Limbs	80	20 bones in each
Ribs	36–38	18–19 pairs
Vertebrae	51–57	7 cervical (neck) vertebrae, 18 thoracic (back) vertebrae, 6 lumbar (loin) vertebrae, 5 fused sacral (croup) vertebrae, and 15–21 caudal (tail) vertebrae
Total	201–209	

EQUIVALENT BODY PARTS, HORSE TO HUMAN

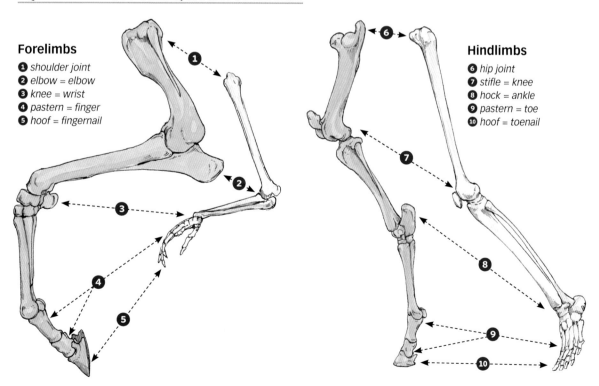

Forelimbs

1. *shoulder joint*
2. *elbow = elbow*
3. *knee = wrist*
4. *pastern = finger*
5. *hoof = fingernail*

Hindlimbs

6. *hip joint*
7. *stifle = knee*
8. *hock = ankle*
9. *pastern = toe*
10. *hoof = toenail*

Horses with short strong backs, short strong loins, and tight coupling tend to be able to carry more weight than average. That's why Icelandic, Arabian, and some Quarter Horses are suited to carry higher weights. A horse in peak condition will be able to support weight better than a thin, poorly conditioned horse. A horse used for walking and posting trot work might be able to carry more weight than a horse that is used for galloping or jumping. But even that depends on the skill of the rider.

A skilled rider sits in balance and moves in harmony with the horse. A loose, crooked, or imbalanced rider continually throws the horse off balance and thus makes his work more difficult. Therefore, a skilled rider might be able to ride a smaller

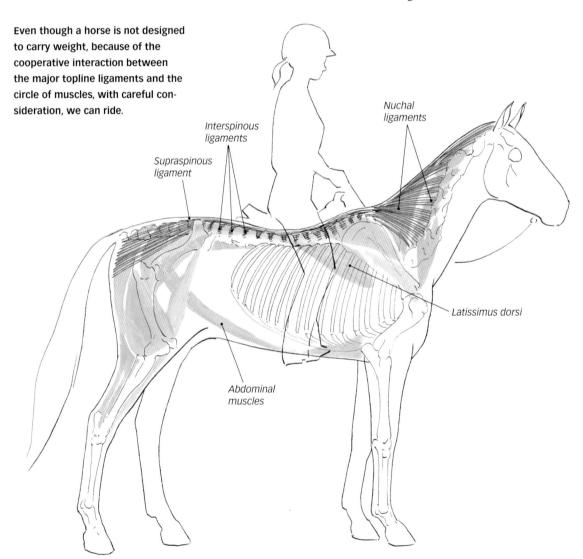

Even though a horse is not designed to carry weight, because of the cooperative interaction between the major topline ligaments and the circle of muscles, with careful consideration, we can ride.

Interspinous ligaments

Nuchal ligaments

Supraspinous ligament

Latissimus dorsi

Abdominal muscles

horse while a novice rider might require a larger, more solid horse to compensate for the erratic movements of the rider.

Finally, the type of saddle can affect the weight-carrying capacity of a horse's back. A rider's weight, as well as the weight of the saddle itself, is distributed on the horse's back by the bearing surface of an English saddle's panels or the tree of a Western saddle. An average English saddle has a bearing surface of about 120 square inches and an average Western saddle has a bearing surface of about 180 inches. So when using a Western saddle, a rider's weight will be borne by an area that is 1½ times the size of the bearing surface of an English saddle. When comparing, you will also need to take into consideration that a Western saddle might weigh 15 to 40 pounds while an English saddle would weigh between 10 and 20 pounds.

Because the back ligaments weaken with age and use, we need to fit saddles well and learn to ride effectively in order to preserve our horses' comfort and usefulness.

Locking Limbs

The horse has unique limb anatomy that allows him to stand while sleeping using a minimum of muscular activity. Three separate features enable a horse to sleep on his feet: the stay apparatus, the reciprocal apparatus, and a locking stifle joint.

The ligaments and tendons of the stay apparatus stabilize all the joints of the forelimbs and the fetlock and pastern joints of the hind limbs so that a horse's weight is supported on straight limbs as he dozes. The reciprocal apparatus of the hind limbs is a set of muscles between the stifle and hock joints that work in unison. When the hock extends, the stifle must extend. When

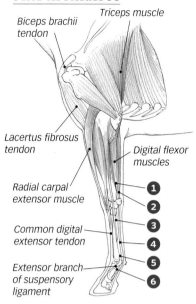

STAY APPARATUS

Labels: Biceps brachii tendon; Triceps muscle; Lacertus fibrosus tendon; Digital flexor muscles; Radial carpal extensor muscle; Common digital extensor tendon; Extensor branch of suspensory ligament; 1; 2; 3; 4; 5; 6

(1) Radial check ligament; (2) Superficial digital flexor tendon; (3) Carpal check ligament; (4) Deep digital flexor tendon; (5) Suspensory ligament; (6) Distal sesamoid ligaments

The stay apparatus supports and stabilizes the horse's forelimbs.

RECIPROCAL APPARATUS AND LOCKING STIFLE

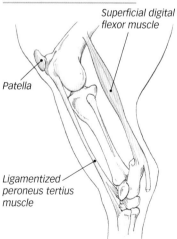

Labels: Superficial digital flexor muscle; Patella; Ligamentized peroneus tertius muscle

The reciprocal apparatus and locking stifle lock the hind limbs during dozing. Although classified as muscles, those identified in this drawing are largely tendinous.

Biostatistics for a 1,000-Pound Horse

Blood volume	9.25 gallons
Stomach capacity	2–4 gallons
Feed in	16 pounds of hay per day
Water in	5–10 gallons per day
Manure production	40 pounds per day
Urine production	6 quarts per day
Mare's milk production (lactation)	9–11 gallons per day of low-fat (1.5%), high-sugar (6.5%) milk

the hock flexes, the stifle must flex. So when the stifle is locked, the hock is locked. The horse's stifle, equivalent to our knee, can be locked when the horse puts most of his weight on that limb. The other limb rests with just the tip of the toe on the ground. Every few minutes, a horse will shift his weight from the left hind to the right hind to alternate which hind rests. (See more about sleeping in Routines, chapter 5.)

Hoof Growth

Horse hooves grow approximately a quarter of an inch per month. Because wild horses move freely and over abrasive terrain, their hooves usually wear off as fast as they grow. Domestic horses, in contrast, often have limited movement and are typically kept on soft footing. Most domestic horses therefore require hoof care every six weeks to trim excess growth and balance their feet. Because the portion of the hoof that the farrier nips or nails is equivalent to the ends of our fingernails and is insensitive, it doesn't hurt a horse to have his hooves trimmed or shod.

A horse with well-shaped, healthy feet that lives and works on ground that is minimally abrasive can remain barefoot. But a horse with low heels, weak hoof walls, thin soles, cracks, or other hoof disorders would most likely benefit from shoeing. Even a horse with ideal hooves would need shoes if worked on ground that wore the hoof away faster than it could grow. My good mare Aria has the best hooves I have seen, yet I keep her shod because I ride in the foothills of the Colorado Rockies.

Equine Vital Signs

LIFE STAGE	TEMPERATURE	PULSE (PER MINUTE)	RESPIRATION (PER MINUTE)
Newborn foal	101	70–100	65
Foal	100	70	35
Adolescent	100	40–60	12–20
Adult	99.5–101.5	30–50	10–14
Senior	99–101.5	30–44	10–15

CHAPTER 4

The Nature of the Horse

ORSES ARE GREGARIOUS. They love company. There is safety and comfort in numbers, and the horse herd represents security. Horses are most content if they can touch other horses, be near them, or at least see them. Nevertheless, unless a horse is soliciting close interaction from a particular preferred associate, most horses like to maintain a zone of 12–15 feet of personal space around themselves.

Many people are that way. I like to be with my family and friends, yet I need my personal space, too. What's different with horses is how they act out their social needs — and this is helpful for us to understand.

Bonding

The first bond, between mare and newborn foal, is very strong. It starts at birth, during the imprinting period, when the mare licks and nickers to the foal and the foal nurses. For the first few weeks, the foal wants to stay very close to its dam and the mare is very protective of her foal.

This bond naturally begins to weaken as the foal matures, so that by six months of age, the foal is more independent and the mare is ready to wean him. In a domestic situation, the foal is usually separated by sight and sound from the mare and paired with other foals or young horses for companionship. In the wild, when a foal is weaned to make room for another foal, the older foal, especially if it's a filly, might remain with the dam and make an even stronger bond.

Alternatively, a weaned foal (wild or domestic) might seek out the companionship of another horse for pair bonding. These new companions become buddies or preferred associates. The bond between them can bring a great sense of confidence and security when they are together, but when they are separated, extreme anxiety can take over. In domestication this is referred to as separation anxiety and is part of the conditions described as herd bound, buddy bound, and barn-sour.

A barn-sour horse suffers separation anxiety and desperately wants to retain contact with a buddy or the barn, which represents security. The herd-bound horse might try to go over or through fences or other obstacles to reach a group of horses. Even a mildly buddy-bound pair of horses can be nerve-racking, because the horses constantly try to keep sight of each other. On a trail ride, for example, just as one horse goes around a bend and into some trees, the horse behind loses sight of him and panics.

An insecure horse might try to retain contact with a buddy vocally, through loud calling. This is an example of an et-epimeletic behavior, meaning behavior that "asks or signals for care or attention." The key to preventing barn-sourness in a horse is to design his experiences so that he learns to live alone at times in his life and also with various horses, not always with one particular horse. In addition, your horse should learn to buddy up with you, instead of with another horse, to derive a similar sense of security.

★   ★   ★

A number of terms describe horses' desire to be together.

Pair bond *refers to the relationship between two horses that exhibit a preference to stay together; sometimes the bond is so strong that it causes problems.*

Separation anxiety *is the nervousness that arises when bonded individuals cannot touch or see each other. This can cause barn-sour, buddy-bound, or herd-bound behaviors.*

Buddy bound *describes a strong bond between two individuals that can result in separation anxiety.*

Barn-sour *describes insecurity that may make it impossible to move a horse away from a barn or may result in his bolting back to the barn.*

Herd bound *describes insecurity that may make it difficult to remove a horse from a herd or may result in his bolting back to the herd.*

★   ★   ★

Dealing with Bonding Problems

Since our horses are home-raised, there are natural maternal bonds between our mares and their offspring. Bonds can become problematic with horses that are constantly housed near each other.

For example, although putting Zipper next to his dam, Zinger, makes for a peaceful housing arrangement, it can have an undesirable effect when Zipper is worked. He keeps one ear and part of his concentration on the whereabouts of his dam. It is as if an invisible umbilical cord has appeared between the 31-year-old mare and her 20-year-old gelding.

The simple management practice of housing them in areas separate from each other diminishes the response by 99 percent. In fact, this past year, after Zipper was given a year off on pasture, he didn't even acknowledge Zinger when she was turned out in the pasture next to him.

A friend's gelding (below) had extreme separation anxiety, often working himself into an unhealthy lather when apart from his buddies.

Assuming that horses and humans can make strong bonds, do horses miss us when we are gone? I think they might miss the feeding and good care we give, but they are more like cats in how much affection they outwardly show. Some horses and some cats are very demonstrative, but most are aloof when it comes to emoting to humans — unlike dogs, which tend to develop very strong emotional bonds with us.

The Single Horse

A horse can adapt to living alone, without the company of other horses. To fill some of your horse's needs for interaction, you can provide him with some socialization substitutes. Regular grooming will help take the place of mutual grooming with another horse. Daily interaction, whether it is ground work or riding, will help satisfy his need for bonding.

If a horse is particularly lonely, you can house or pasture him with a companion animal such as a small pony, burro, goat, or other farm or domestic animal that works for your situation and your horse.

Mutual Grooming

Mutual grooming is an example of epimeletic behavior, giving care or attention to another. The first instance of epimeletic behavior is when a mare licks her newborn foal. Mutual grooming comes later and is the equivalent of "you scratch my back, and I'll scratch yours." It usually takes place between two preferred associates, because a horse has to trust another horse to let him nibble along his neck, withers, and back!

★　　★　　★

Epimeletic behavior *means giving care or attention.*
Et-epimeletic behavior *means soliciting care or attention.*

★　　★　　★

When you first groom your young horse on the withers, it would be natural and instinctive for him to turn and try to reciprocate, but it would be pretty painful if he nibbled you with his incisors. You can say "Thanks, but no thanks" by tying him in a fashion that prevents him from turning around and reaching you while you are grooming. Eventually, his instinct to reciprocate will diminish.

In addition to nibbling, mutual-grooming partners often stand head to tail and swat flies off each other.

Most horses enjoy a vigorous session of mutual grooming with their preferred partner.

Pecking Order

A pecking order is the dominance hierarchy that exists in every group of animals, from chickens to dogs to elephants. Horses, whether in a pair or a large herd, rank themselves in order of authority. Factors affecting a horse's position include: age, body size, strength, athletic ability, sex, temperament, and the length of time in a particular band. When the pecking order is being established among a group of horses, it can often be quite violent, with kicking, biting, and chasing. Once it is established, however, future aggression is unnecessary because every horse knows his place.

Humans occupy a position in the horses' pecking order, as well, so we must convince them not only that we are the top horse, but also that we are wise and fair leaders. For safety and smooth management, a human *must* be the top horse.

Pecking order is most evident at feeding time. You can easily tell a human's rank among horses by watching as she feeds them. If the horses come charging into the person's space and she drops the feed and turns tail, one of the horses is definitely on top. In a pasture group, things can become pretty active, but pecking order activities take place between stalled and penned horses, too.

Each time a new horse is introduced into a herd, a band, or a barn, things will mix up a bit. That's why, to prevent injury, it is important to introduce new horses gradually. Put the new

A group of horses can be violent while sorting out its pecking order or social rank, but eventually settles into an orderly system.

horse near but not directly across a fence from the existing group for several days so they can see and smell each other. After a few days, turn the new horse out with some of the group members that are the most agreeable and then gradually add others back to the group.

The Battle of the Sexes

Mares rule! A horse herd is like a clique, gang, or group of friends that is run by women. In the wild, horses live in a matriarchal society, and this order often carries over to domestication, as well. Matriarchs are also called lead mares, boss mares, or alpha mares. In addition, mares usually form pair bonds for mutual grooming and running games.

Zinger, my 31-year-old mare, is the undisputed matriarch of our ranch. She receives preferential treatment and respect from all. All horses note when she is being moved around the property. When she is in a pasture band, she is the acknowledged leader and never has to fight for her position. She initiates the exodus down to the creek to drink. When she finishes grazing and finds a place to rest, all of the other horses in the band rest near her. She is no pushover when it comes to goofy youngsters, mares in heat, or geldings that don't know their place.

Gelding groups are the domestic equivalent of bachelor bands in the wild. They tend to show off, engaging in mock fights and extreme races. Many geldings don't mix well with mares, but there are exceptions. Zipper is a neutral gelding; he rarely oversteps another horse's boundaries, nor invades a mare's critical distance, nor does too much sniffing, so he can safely be turned out with mares and other geldings. Dickens, on the other hand, is trouble looking for a place to happen. Turning him out with or near mares always results in a lot of squealing and urinating, running, herding, biting, and kicking, so Dickens is often turned out alone or with other geldings. In fact, because he likes to touch, sniff, and nibble other horses, he makes a good "teaser" — a horse that lets us know when a mare is receptive and in heat for breeding.

Is this a mock fight between two playful geldings, or a real battle for pecking order position or sexual dominance? One horse goes for the jugular while the other defends with his forelegs. (Let's hope it is just play.)

Stallions in the wild play a very different role than domestic stallions do. Their job is to keep their mare bands together and pregnant. Domestic stallions, on the other hand, are primarily bred using in-hand breeding or artificial insemination; consequently, they live in solitary confinement, never enjoying or benefiting from natural interaction with herd members. Turning a stallion out with geldings, even if they aren't a threat sexually, often results in fighting and injury. Exceptions to this are the stallions that live on large ranches in the West. It is somewhat amazing how quiet these naturally socialized stallions can be. Except for their obvious stallion physique, they could be mistaken for geldings.

Castration

Castration is a common procedure for 90 percent of domestic male horses. Geldings generally have a more stable disposition than do either stallions or mares and are well suited for a wide range of uses. In contrast, vocalization, fractious behavior, and sexual interest in fillies and mares are typical characteristics of the yearling stallion. Although gelding, by itself, will not change established behaviors, such as nipping and teasing, it will eventually remove the tendencies toward them.

Gelding at one year of age is common. Waiting longer may increase muscle development but also establish stallion behaviors more deeply. A gelding will not be able to impregnate a female horse, although he may retain some stallionlike behavior for a time.

Horse Play

Play among horses can be rough, yet it is essential behavior for both the physical and social development of foals and young horses. Playing includes running (alone, with one other horse, or with many others), chasing, bucking, rearing, jumping, nipping, biting, striking, and kicking. The first game a foal learns is running, then rearing, then a favorite of all ages: approach, withdraw, turn and kick.

Mares tolerate a foal's play but rarely enter into the playing. It is best if young horses are allowed to play with horses their own age, anyway. Older horses aren't interested in the same games and many are impatient with a foal's vitality. During times of stress, such as extreme weather, drought, or scarce feed, playing decreases greatly.

Play teaches a foal fighting behavior, sets the stage for sexual training, sharpens reflexes, develops the competitive spirit, and improves overall stamina. Foals begin to test their limits from day one and continue these antics as long as they have an

Foals, especially colts, will often play-box while standing on their hind legs. This is an excellent example of why horses shouldn't wear halters on pasture. Can't you just see a forehoof caught in a playmate's halter?

interested playmate. Colts invite play by nipping or nudging another horse on the leg, then both horses drop to their knees and spar and bite. They even play-box on their hind legs, biting and striking at each other with their forelegs. Fillies don't play as roughly as colts. They mainly run, kick, and participate in mutual grooming.

Many horses pick up objects with their mouths and play with them. These can include sticks, tack, clothing, feed sacks, and ropes. After grasping the object in their teeth, they often shake and fling it around.

If a young horse has had inadequate exercise and socialization, he may attempt to play with you during lessons. Since horses play rough, it is best to discourage your horse from playing with you and provide an opportunity for him to play with another horse.

Many horses enjoy picking things up with their teeth and tossing them around.

Curiosity or the Investigative Behavior

One of the most endearing and valued characteristics of the horse is his curiosity. Strong investigative behavior is what helps a foal get his first drink of mare's milk and survive.

Take great care not to discourage a horse's curiosity, because it is an essential key to his behavioral development and a means for him to learn about things. Plus, it is a valuable way to help you train the horse. Curiosity shows interest, a precious gift.

Horses have remarkable curiosity, and this is a valuable trait that we must be careful to preserve. Suckling Drifter and his sire Drifty decide that the sight, smell, and sound of the mower aren't so threatening, so they come up to take a closer look and sniff. Allowing your horse a chance to satisfy his curiosity will result in a more confident animal.

When a horse sees something new or unusual he goes through a sequence of reactions, as follows:

1 His first reaction is suspicion, with an alert stance and adequate distance from the object.

2 As he begins to take in information about the object via his senses of sight, smell, and hearing, he begins moving in closer, often in a circling fashion.

3 When he closes the gap enough to touch the item, he might find it is safe and beneficial, like feed or water, or he might decide it is dangerous and put more distance between it and himself.

An interesting experiment is to walk out to a horse pasture and just squat down and be still. The reaction among horses will vary, but almost invariably, after the initial snorting and sashaying back and forth, one brave soul will come up for a sniff. If you move suddenly during the critical distance phase, however, off he'll go.

Curiosity can also get some horses in trouble when they learn how to operate latches, gates, faucets, waterers, light switches, and the like.

The Nomadic Lifestyle

Horses are born wanderers. For wild horses, the distance between feeding or watering areas often required extensive travel, so they survived by eating many small meals while continually on the move. With muscles in motion and warmed up, nerves ready to fire and senses keen, the horse in motion was successful in detecting and escaping predators.

Most horses are basically followers. The decisions of where to go and when are made by only a few individuals, most often the lead mare. Domestic horses retain these instincts. They want the freedom to roam and to follow a trusted leader.

Follow the Leader

The horse is content to be a follower when he finds a strong and just leader. You can use this to your advantage in training. First of all, if you establish a connection of mutual trust and respect with your horse, he will be willing to follow you. Additionally, when introducing an inexperienced horse to a new obstacle, you can have a friend lead an experienced horse over or by the obstacle, in front of the younger horse. This can be especially handy when crossing a strange creek or stream. If he trusts you and considers you his leader, you can also dismount and lead him across an obstacle. (See page 133.)

Horse trainer clinicians use the horse's tendency to follow with dramatic effect when they induce an unfamiliar horse to connect with them and follow them around without a halter or lead rope.

Most horses are followers and prefer that a trusted leader make all decisions.

Routines

T HE HORSE IS A CREATURE OF HABIT. He is most content when he can order his day according to his instincts and needs.

Although wild horses have been observed to stick rather closely to definite daily patterns, seasonal variations occur, as do daily changes determined by the weather, temperature, air flow, humidity, and atmospheric pressure. Horses tend to be restless in the wind, lethargic in high humidity, and unpredictable when the barometer is changing. In spite of all of the many variables, however, horse routines are fairly consistent.

The Horse's Biological Clock

Horses improve our lives. Their strong biological clocks, seasonal changes, and daily routines bring a sense of order to a sometimes chaotic human calendar.

Eating

The most important routine from the horse's point of view is eating. Since horses evolved as nomadic grazers, their digestive systems are set up for small, frequent meals. Pastured horses generally graze for sixteen hours a day. Confined horses look forward to having their meals on time and can become noisy and upset if feedings are missed or late.

Free-choice feeding doesn't work as well with domestic horses as it does with chickens and some dogs and cats. Horses tend to overeat. An exception might be feeding free-choice grass hay, but often this results in wasted hay that is trampled and soiled and then left uneaten.

Drinking

It's really true: You can lead a horse to water, but you can't make him drink. Horses drink only when they want to, and, typically, that is soon after eating the bulk of their roughage.

Horses that don't have access to water can die of dehydration or impaction colic. Having fresh water available at all times is the best insurance that a horse will be able to satisfy his thirst when he is ready.

Defecating and Urinating

Many horses, whether in pasture or in close confinement, choose particular places to deposit their feces. With stallions, it is often part of a scenting ritual and territorial marking. With other horses that are naturally neat, it is probably just a happy coincidence, so be thankful that they are so inclined, as it makes collecting the manure much easier. In large pastures, the horses usually designate separate areas for eating and for elimination. This is a natural means of parasite control.

Horses defecate every two to three hours, from five to twelve times per day, and more frequently when stressed or ill. Each bowel movement consists of five to twenty fecal balls. Each fecal ball can contain as many as thirty thousand parasite

Top on a horse's to-do list is to get enough to eat. This can take as much as 16 hours a day.

When you gotta go, you gotta go. Most horses wouldn't urinate on a hard roadway because of splash back, but this police horse has adapted to his working environment by adopting an extra-wide stance.

Mares' Ways

Urination can be much more frequent with a mare in heat: she will frequently squirt tiny amounts. Horses learn to urinate by signals. My mare Aria almost always urinates in her gravel pen when she sees me heading to the barn at feeding time. Then she enters her matted feeding area for a relaxed bout of chewing her grass hay.

eggs! Re-infestation with parasites will occur daily if horses are forced to eat in an area of fecal contamination. Therefore, it is wise to offer sanitary feeding areas.

Although horses (but not some ponies) can defecate on the move, only needing to raise their tails, urinating is a more ritualistic procedure. It begins with choosing a place to urinate that won't splash back, because horses don't like to splatter their legs. The stance is different for male and female horses. With mares, the head is low, the back is arched, the hind legs are extended back and spread apart, and the tail is raised. Male horses stretch out, separating the forelegs from the hind legs more than mares do, which makes their back flat or a bit hollow.

Horses urinate about every four to six hours. When you are on a long trail ride or your horse is on a long trailer trip, be sure to allow him opportunities to relieve himself. When traveling, the horse may not urinate on board if the trailer is not bedded, so he will need to be unloaded periodically to urinate in the grass. If you are mounted and your horse starts to assume the stance, rise up off his back.

Caution: if your normally good horse seems unusually antsy while being groomed or shod, it could be that he needs to urinate but doesn't want to splash on the barn floor. In this case, discipline obviously won't solve the problem. Instead, turning him into a paddock or freshly bedded stall for a few minutes should do the trick.

Almost any horse will urinate in a freshly bedded stall. It is thought that they want to scent it and make it seem like home. That's why when racetrack officials need a urine sample from winners of a recent race, they put them in stalls with deep straw bedding. The horses can't resist giving a sample!

Grooming

Horses perform various grooming rituals on a regular basis. Besides mutual grooming, they groom themselves. Self-grooming includes rolling, rubbing, scratching, and nibbling, and is heightened in intensity during shedding seasons and muddy periods.

Horses roll on the ground to scratch themselves, to remove loose hair, to counteract the plastering effect of rain, to loosen sweat, and to coat themselves with a layer of dirt or mud to

ward off insects. Prior to rolling, a horse may loosen and soften the soil by pawing. After he rolls, he stands up and shakes off his coat. A horse should be allowed to fulfill this natural desire. To make grooming easier, you can control where he rolls, such as by turning him out after a lesson for half an hour in a sand pen. This will allow him to perform his ritual at the same time that he cools and dries from his workout.

Rubbing is another favorite pastime of horses. It helps them scratch an itch or insect bite; removes hair, mud, and sweat; and often relieves the discomfort of a wound. Horses rub themselves with their teeth (biting their forelegs and sides or blankets) and their hind legs (scratching their heads and necks with the toe of a hind hoof). They also rub any and all parts of their bodies on buildings, fences, and trees. This can rip blankets, turn a wound into hamburger, or make a bald spot at the top of the horse's tail. To a horse, though, all of this is normal self-grooming.

For safety, it is important that a horse not be turned out with a halter, blanket, or other item of tack or clothing that could trap him if caught on a fence, tree, or his own hoof. Breakaway features are available for items, specifically halters, that are used for turnout.

Wearing a blanket does not diminish the desire for a good roll. *Note:* all blankets and pasture halters, if used, must be outfitted with breakaway safety features.

How Horses Lie Down and Stand Up

A horse lies down to roll or sleep in a prescribed way. First he steps forward with his hind legs and back with his front legs so that all four feet are close together, as if he was practicing for the circus. Then he might rotate until he finds a good place or orientation to lie down. When satisfied, he will raise his head, flex his forelegs, kneel onto the front of his pasterns and knees, then tuck his hind legs under his body and lower himself so that his belly is resting on the ground.

When the horse stands up, he extends his forelegs out in front of his body, raises his forehand, and then uses his hind legs to lift his hindquarters.

If a confined horse has only hard or abrasive ground on which to lie, he can develop bleeding sores on his legs. That's why he needs a soft place to lie.

LYING DOWN

Watch the horse's forelegs when he lies down or stands up. When lying down, he (1) flexes his forelegs, (2) kneels, and (3) lowers his hindquarters down.

RISING

When rising, he (4) extends his forelegs, (5) lifts his forehand, and then (6) lifts his hindquarters.

1

4

2

5

3

6

Sleep Positions

Horses rest in one of three sleep positions for a total of five to seven hours in a twenty-four-hour period. The more secure and relaxed a horse is, the more likely he will rest in positions 2 and 3.

1 Dozing. Horses doze while they are standing with lowered head, the posture of minimal energy demand on the horse. Their stay apparatus and reciprocal apparatus allow them to lock their front legs and rest one hind leg at a time while they doze. Their eyelids can be in any position from half-open to almost closed. Yearlings and older horses may spend a total of four resting hours each day in this position. The slow-wave sleep of dozing is characterized by decreased heart and respiration rates and decreased muscle tone.

Position 1: Dozing while standing

2 Sternal-Recumbent. When a horse lies down for a sternal-recumbent nap, he lies on the midline of his belly, sometimes resting his chin on the ground. His legs are usually folded underneath his body and he can rise in an instant. Slow-wave sleep generally occurs when the horse is in the sternal-recumbent position. About two hours per day are spent in this position. Any longer and the compression of the organs becomes uncomfortable and could be harmful.

Position 2: Sternal-recumbent

3 Lateral-Recumbent. From the sternal-recumbent position, the relaxed horse might roll over on one side, lay his head and neck flat on the ground, and stretch his legs out in the lateral-recumbent position. The horse's eyes will be fully closed. When he is lateral-recumbent, he experiences deep REM sleep, so you might see him twitch or "run" with his legs, or hear him grunt, snore, or whinny. His respiration and heart rate are slightly elevated. If environmental conditions allow,

Position 3: Lateral-recumbent

a horse will sleep in this manner for up to one hour per day.

Precocial species, like the horse, tend to have lower REM sleep requirements than do animals that are helpless at birth, like humans. A horse in REM sleep is hard to rouse, so he will only stretch out like this if he feels very secure.

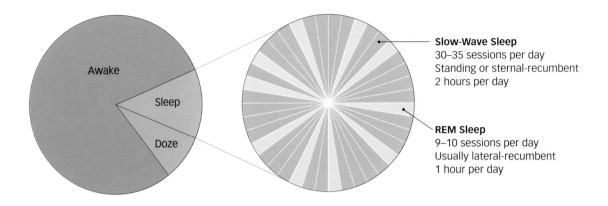

Slow-Wave Sleep
30–35 sessions per day
Standing or sternal-recumbent
2 hours per day

REM Sleep
9–10 sessions per day
Usually lateral-recumbent
1 hour per day

Awake

Sleep

Doze

Resting and Sleeping

Humans tend to do all their sleeping in one period at night. Horses sleep twenty to fifty times a day in tiny naps. Wild horses do all of their sleeping during the day, in very brief spells, and keep vigil for predators at night. Domestic horses sleep during both the day and night but retain the instinct to be vigilant by mainly dozing or sleeping in short bouts.

Foals, especially during their first few weeks, spend more than half of their time sleeping and the rest nursing. The healthy foal will most commonly be observed resting in the sternal-recumbent or lateral-recumbent position (see page 71). By the time a horse is two years old, he has formed a close approximation to adult resting habits.

Shelter

It is surprising how hardy horses are. While I recommend providing natural or man-made shelter for horses, you'll often see a horse standing outside a shed in the midst of a terrible winter storm. I can only imagine that such horses feel confined and a bit trapped when they are inside, and given their druthers, when there is a storm threatening, they'd rather be outside, free and ready to flee. Domestic horses that refuse to use a run-in shelter during a storm might dislike the drumming of rain or hail on the roof or remember the sudden, frightening sound of snow sliding off the roof. Life is safer and simpler in the wide-open spaces. A horse's coat is well suited to keep him comfortable in many kinds of weather.

Self-Preservation

I admire a horse with a high self-preservation aptitude, one that avoids getting hurt or cornered. Such a horse is alert and takes care of himself and is generally less costly to care for. Some horses, like my mare Zinger, who has had only one injury in thirty-one years, avoid situations that are risky. Others, like Dickens, are more lackadaisical and goofy and are constantly getting cuts and scrapes.

When a horse notices a new object in an otherwise familiar place, he might pause, startle, or spook. For example, I rarely drop a horse blanket on the floor of the barn, but if I did so and then led a horse into the barn, even though he might be very familiar with blankets, he would most likely stop momentarily (mentally and/or physically) to decide if the blanket poses a threat: "What the heck IS that?!"

Dogs

Mares usually foal at night to reduce the chances of a predator killing the foal or the mare while she is in labor. This also allows the foal a few hours before dawn to gain strength and figure out how his legs work. It is understandable, then, why many domestic horses vehemently dislike dogs. They might flee from them, right through fences at times, or they might turn and chase one with their teeth bared.

Even my 29-year-old mare Sassy, whom I raised from a foal and who has lived with dogs all her life, can still turn quite aggressive toward dogs. She's not fooling around — but then,

One type of shelter that horses do take advantage of is shade. Whether cast by trees or sheds, shade allows a horse to escape the hot summer sun and avoid insects that prefer sunny areas.

These buddies didn't read the prey-predator manual and thoroughly enjoy each other's company.

she has backed down some ornery cows and the odd llama or emu, too. She has a very strong sense of self-preservation and a protective maternal instinct, but no fear.

I generally exclude dogs from training areas unless a particular horse and dog know each other well or have an affinity for each other. Bonding can happen between horses and dogs, too. To every rule, there are exceptions.

Flight — Survival — Spooking

When something scares a horse and he can flee, he will run first and ask questions later. He will put a certain distance (called flight distance) between himself and the fearsome object or situation, then stop and look back at it. The flight distance could be a few feet from a shirt on the clothesline, or a quarter of a mile and still running from a bear. When the horse finally does stop to look and the object is no longer there, he actually becomes more fearful than if he could still see it.

You might say that horses have great imaginations when it comes to danger. Real or imagined — it is all the same to them. Even when a horse is not particularly afraid of something, he still operates with a flight zone around him, a particular area of personal space that only trusted associates (human or animal) or familiar things are allowed to enter.

Spooking in Place vs. Running

A seasoned horse might have built up enough courage and a big enough repertoire of strange experiences in his memory bank that when he encounters something odd, he is simply startled. The startle response is also known as a spook in place. His heart and body jump but his feet stay still. We can certainly live with that.

When a horse shies, however, he often spooks and runs. This can be dangerous if you are riding him.

Things That Worry Horses

Although pheasants and ducks that take sudden flight and deer that bound out of the bushes don't pose a deadly threat to a horse, their sudden movement and sound can catch him by surprise and cause him to jump sideways very quickly. When this happens, even a seasoned horse and rider can come unglued.

One fine spring day, Zipper (this gelding has appeared in my books for years; he was seventeen years old on this occasion) and I were enjoying a rein-swinging walk across a pasture we'd ridden through many times before. On this day, however, there was a fawn hidden in the tall grass. When we came upon it, rather than step on it, Zipper jumped up and twenty feet to the left, landed, and just stood there, looking at the spot where the fawn was. Meanwhile, I hovered in the air without a horse under me for a while until I finally had to hit the ground.

Also scary is almost any common object that is out of place, that moves in a sudden or strange manner, or that makes an unusual noise. Examples are an umbrella opening, motorcycles and bicycles, plastic bags and sheets, balloons, water puddles, streams, and rivers.

Petrified of Pigs

Even though a domestic pig probably doesn't represent a threat to horses (although perhaps a wild boar did to eohippus), the odd behavior and smell of pigs, and their sudden noise and quick movements, alarm many horses.

Zipper Stays Zipped

When the phone company was dynamiting through rock to install underground lines along our road, the horses never really got used to the loud, unpredictable sounds, even though the blasting crew worked near us for several weeks. The erratic blasts bothered me, too, but I reasoned. There would be a blast, my heart would jump; I'd go on.

One day, I was crossing Long Tail Spring on Zipper just as a blast went off behind us. Before the blast, he was relaxed, moseying along, with his head and neck long and low as we crossed the water. In an instant, when the dynamite ignited, his body shortened; he tucked his tail, dropped his croup, and raised his head. He became tense and hard and his entire body shot straight upward. When he landed, all four feet splashed in the spring at once, sending mud and water up at us both. Then he just stood there as if to say, "What the heck was that?" When I urged him on, he continued forward steadily but alertly. His reaction was understandably intense. In a similar situation, it would not be unusual for a horse to buck and bolt out of fear.

CHAPTER

6

Good Behavior, "Bad" Behavior

To a good horseman, and to the animal himself, everything a horse does is just behavior. When we label something a horse does as bad, we are basically saying, "I don't like it; I don't understand it; I don't know how to change it."

The more we know about the nature of the horse and how domestication and training affect him, the better chance we have of helping all of our horses be straight-A students.

To learn, observe, and enact takes time, but how better to spend your time? Content horse — happy, safe rider.

The Spirit of the Horse

Up to this point, I've been talking about specific physical and behavioral characteristics. It is the horse's spirit, however, that makes him such a trainable partner for us lucky humans.

Here are some things I've observed:

★ *Horses are cooperative and willing.* They are generous and forgiving and make patient teachers. They have the precious trait of curiosity, which we should take great pains to preserve. They are trusting, and once we have earned their trust, we should maintain it.

★ *Horses are adaptable* to a variety of situations, are good learners, and have an excellent memory.

★ *Horses are not naturally aggressive* and can quickly learn to be submissive, so much so that others often take advantage of them.

★ *Horses have a special sense* that allows them to detect our moods or a shift in the weather before it occurs. They seem to know when there is "something in the air," whether it is something playful or impending doom. Without physical contact, they can read and pick up very subtle signals from a human. That is what makes horses so valuable for the rehabilitation of people who have suffered physical, emotional, or mental damage.

A horse's spirit is strong, yet sensitive.

Temperament and Attitude

Temperament is disposition, the general consistency with which a horse behaves. It is a steady, permanent characteristic of a specific horse. We have little, if any, effect on a horse's overall temperament.

Attitude, on the other hand, is temporary. It is the outlook of a horse at a particular time and may be affected by specific conditions, such as hormones, the weather, recent events, good or bad memories, or illness. A good horse might temporarily have a bad attitude, and a spoiled horse might have a good day every now and then. Attitude is something on which we humans can have an effect.

★ ★ ★

When discussing temperament, horsemen use specific terms. A **sullen** *horse, for example, is one that is sulky, resentful, and withdrawn.*

★ ★ ★

Horses Are Individuals

Even though most horses show certain universal behavior characteristics, every horse is an individual with his own disposition. Many factors contribute to a horse's temperament. Some are inherent at birth, while others result from early experiences and the environment. Genetic influences include type, breed, and family lines.

Influences on Temperament

Whether a horse is more cold-blooded (draft) or hot-blooded (Thoroughbred and Arabian) depends on his breeding. Ancestry affects temperament factors such as sensitivity, athletic potential, and level headedness, as well as physical characteristics like hair coat, bone size, thickness of skin, density of hooves, and size of eyes and nostrils. Some of these conformational traits can have a strong influence on temperament.

Temperament Types

Every horse is unique and has his own temperament. There tend to be some common threads among horses, however, that allow us to categorize them by temperament. Profiling, by nature, is gross generalization, but it can provide a handy way to talk about a horse in broad terms. Most horses are a combination of two or more types.

Physical characteristics (such as the size and set of the eyes, hair whorls, or thickness of the skin) may or may not indicate the temperament a horse is likely to have. All horses are trainable, but some take to it more easily due to their temperament.

Alert. Interested, kind, and cooperative. Pays attention and likes to learn and interact with people. Generally calm and confident, yet responsive. In my experience, this describes the vast majority of horses.

Stubborn. Might be dull or lazy; can be tough to work with because when this horse doesn't understand or gets tired, he can become sullen and tune out mentally, turning off the message center connected to his senses. Generally requires more time and patience to train than an alert horse does and can require continual retraining.

Nervous. Hyper, easily excitable, and hard to calm down. Tends to shy and spook. Has a lack of confidence, which may be improved. Can over-anticipate. Would resent and react, often violently, to restriction and restraint. Best trained by an experienced horseman. Some mature beyond the nervousness and are trained into very responsive, excellent horses.

Aggressive. Characterized by explosive behavior, such as charging, biting, kicking, and striking. Seldom can be safely trained, handled, or ridden except by rare accomplished horsemen.

The physiology of the senses, for example, can contribute to a horse's confidence. Horses with large, prominent eyes have a wider field of vision than do those with small pig eyes. Certain breeders select family lines that have physical traits and a temperament that result in a more trainable horse.

The sex of a horse and the subsequent hormone levels affect behavior patterns. Stallions tend to be more aggressive, with a strong drive. Mares, with their hormone changes, can be unpredictable and sometimes as aggressive as stallions.

Other physical factors that affect temperament are age, health, physical condition, and diet. Of these, all but age are governed by the environment of the wild horse and by the management of the domestic horse.

Early experiences with humans and with other horses have a lasting effect on the personality of a young horse. Horses need to learn acceptable limits of social behavior within the herd hierarchy. If they are sheltered from interaction when young, they often do not act wisely when it later becomes necessary to socialize with other horses. Letting a horse be a horse is important for his social development.

The vast majority of horses are alert, kind, and cooperative.

Natural Horsekeeping

Keeping horses is inherently unnatural, but we can all design our facilities and management routines to give horses as natural a life as our situation allows. The best way to prevent vices is to house and care for your horse using the principles of natural horsekeeping. These include:

★ As much turnout as practical in an area where cantering is possible

★ Living with companions or near companions

★ A large pen with sheltered loafing and eating areas

★ Free-choice grass hay, or a minimum of three feedings per day of grass hay

★ Minimal grain

★ Free-choice salt, minerals, and freshly drawn or naturally aerated water

A stall should be a cozy, restful place, not a prison. Claustrophobia can be reduced with Dutch doors that allow a horse to see other horses and activities.

Domestication Pressures

Many of today's horses are often subjected to extended confinement and a demand to adapt to the schedule of humans. They are overfed concentrates and underfed roughage, and are underexercised, confined, and isolated.

Those who don't fully understand the nature of the horse are more apt to subject him to conditions that are workable for the human but are bad for the animal. For example, horses are basically claustrophobic. Stall life can be so stressful for a horse that he has difficulty being comfortable mentally and physically. What might seem like a cozy stall to you might be a jail cell to him. And what you might consider a healthy routine to keep your horse clean (weekly bath, close clipping, and blanketing) might actually be leaving him without his natural protective sebum, guard hairs, and the chance to have a good roll and be a horse. It is a remarkable testament to the adaptability of the horse that so many learn to love their domestic digs and our care.

Stereotypies

Horses that can't adapt to their living conditions often develop behavior abnormalities, or stereotypies. Some horses, by temperament or heredity, are more prone than others to develop stereotypies. Stereotypies are often signs that a horse is trying to cope with conflict, uncertainty, or restriction.

Conflict occurs when the horse has two opposing urges, both equally strong. A new horse wants to eat but is afraid to enter the barn to get his feed because he is afraid of confinement and of the people in the barn. He might run in, grab a bite, and then dash out, hitting his hip on the doorway. Pain confirms his fear, but he is getting hungrier. He ends up pacing and pawing outside the barn door.

Another example of conflict is the horse that has received a shock from an electric waterer. The owner repairs the waterer in front of the horse and then tells the horse it is now OK to drink. The horse does not understand what the person has done or said, so he does not know that it is safe to drink the water. He sees the water and wants to drink but has a fear of being shocked. He might paw at the waterer (possibly breaking it) and whinny, soliciting attention.

Blue's Blues

I once bought a beautiful blue roan filly off a range in Wyoming, where she and all of her ancestors had been raised roaming thousands of acres. Although Blue had excellent conformation and was a fine student, she had trouble adapting to confinement. In the two years I owned her, I turned her out on pasture more than any horse I had and as much as our land would allow. There were times, though, when the pastures had to rest and I had to house her in a large pen like all of my other horses (who, by the way, were as happy as clams under the same management).

At our place, each horse has his own pen and all are roomy, but some are larger than others. Blue had the largest pen of all the horses. Even with daily work sessions and exercise, however, she developed two odd behaviors that eventually prompted me to sell her so she could return to the open range and become a broodmare.

Blue's stereotypies were unique and they intensified the longer she had to live in a pen. The first was odd and noisy but not particularly damaging. She would hang her head over the top of the panel of her pen and bob her head up and down rapidly and with great force, not just once or twice, but for hours. This made a racket as the panels clattered, and the obsessive behavior wore a bald spot on the underside of her neck. Even putting her in a pen made of the tallest panels on the market at the time (5'6") didn't deter her.

If that was the only aberrant behavior she had, however, we could have engineered a special pen for her. It was really number 2 that worried me and caused me concern.

Every morning when we got up to do chores, we wondered what configuration we would find Blue in that day. She started out lying right next to the panels and putting her legs outside the pen. We blocked the base of the pen with railroad ties and granite rocks to prevent her from being able to stick her legs out. Then she somehow managed to get her legs out *above* the ties and rocks, so that her body was quite a bit lower than her legs. Even when all of the spaces were blocked so that she couldn't get her legs out at all, she would still roll along the side of a pen and become cast. Later, she seemed to be cast in the *middle* of a pen. Often there would be two or three bowel movements under her tail where she lay. A sight strange enough to strike fear in the heart of any horse owner!

No matter how or where we housed her, Blue continually cast herself, often requiring us to dismantle pens or other facilities to free her. Although this made her a constant management concern, my real worry was her health, because lying down for more than a couple of hours can be very dangerous for a horse. For this reason alone, I reluctantly sold her, but I made sure it was to someone who had enough land for her to be out in a broodmare band. I'm glad to report that years later, she is still happy in her range life, and she has raised wonderful foals each year for her new owner.

Uncertainty results when the horse faces a problem beyond his power of resolution. When a young horse is rushed in his saddle training and is asked to do something that he does not understand, he lacks confidence and is uncertain of the outcome. In trying to cope, he might try bucking, bolting, rooting at the bit, or just becoming sullen and tuning out. For example, if a horse is asked to perform flying lead changes before he has learned to lope on the correct lead in both directions, he doesn't know what to do and he doesn't know what the trainer might do.

When a horse is trained progressively, on the other hand, he becomes more confident with each step. He knows what is being asked of him and what the outcome of his behavior will be. When he can't process the requests and cues because he doesn't have the basics, he is uncertain. This could cause him to come unglued.

Restriction and **restraint** are related but are significantly different. Restraint is generally the term used to describe limitation of movement by tack; bad habits such as rearing, bolting, and bucking can develop when restraint is used improperly in handling and riding. Restriction relates to living quarters, and occurs when a horse's movement is limited by confinement. A horse that is restricted to a 12' × 12' box stall six days a week and only ridden on Sunday can pace, paw, and kick his stall and turn into a holy terror in the arena or on the trail. He has a week's worth of energy pent up and wants to get out and exercise. He might buck, bolt, or rear.

Vices and Bad Habits

When a horse develops a behavior abnormality, it is generally one of two types: a vice or a bad habit.

Vices, which are reactions to life in confinement, include wood chewing, pawing, tail rubbing, weaving, pacing, and kicking the stall. Most vices are coping mechanisms. They are best prevented and treated by attention to proper diet, exercise, and socialization. Boredom and the resulting vices are symptoms of inadequate management.

★　　★　　★

Vices are undesirable behavior patterns that emerge as a result of domestication, confinement, or improper management.

Bad habits are undesirable behaviors that develop in response to handling or riding.

★　　★　　★

Bad habits are responses to improper handling and training. Rearing, pulling, biting, and striking are undesirable behaviors exhibited by the horse who feels rushed, threatened, or confused in his training.

Pawing is one way horses react to confinement.

Dealing with Abnormal Behavior

For all vices and bad habits, first examine your management and handling techniques. Modify routines according to the suggestions given here and in the book *Horsekeeping on a Small Acreage,* second edition. If, after concerted efforts on your part, no positive changes in behavior occur, consider alternative restraint remedies that are appropriate to the problem, mechanical or electronic devices, or, upon consultation with your veterinarian, medications or surgery.

If a vice or bad habit makes a horse dangerous or unusable, as a last resort to save his life, you may need to consider these other solutions (see charts on pages 84–89).

Vices at a Glance

VICE	DESCRIPTION
Becoming cast (not a true vice, but an undesirable stable behavior)	Chronically rolling near stall or pen wall and getting stuck alongside or under wall or panel.
Bedding eating	Eating straw or sawdust.
Blanket chewing	Chewing or tearing blankets and sheets.
Bolting feed	Gulping feed without chewing.
Cribbing	Anchoring of incisors on edge of an object (post, stall ledge), arching neck. Can cause colic, poor keeper (prefers "mind drugs" over food); can be socially contagious.
Kicking other horses (not in play)	When turned out with other horses, kicks for any or no reason.
Masturbation	Various methods of self-stimulation and ejaculation in a stallion.
Pawing	Digging holes; tipping over feeders and waterers; catching leg in fence; wearing hooves away, losing shoes. Most often a vice of young horses.
Self-mutilation	Biting flanks, front legs, chest, scrotal area with squealing, pawing, and kicking out. Onset at 2 years; primarily stallions, some geldings.
Stall kicking	Smashing stall walls and doors with hind hooves, resulting in facilities damage and hoof and leg injuries.
Tail rubbing	Rhythmically swaying the rear against a fence, stall wall, trailer butt bar, or building.
Weaving/pacing/ stall walking	Swaying back and forth, often by stall door or pen gate/Repeatedly walking a path back and forth.
Wood chewing	Gnawing of wood fences, feeders, stall walls; up to 3 pounds of wood per day.

CAUSES	TREATMENT
Normal causes of rolling: shedding, blanket fit, colic.	**Manageable.** Serious if horse left for long period of time unnoticed, as he can develop severe colic. Bank stall bedding against stall wall; use antiroller (surcinglelike item) in conjunction with sheet or body blanket. Provide horse with open place to roll on a regular basis.
Compulsive eater.	**Manageable.** Be sure horse gets adequate roughage in the form of long-stem hay. Use nonpalatable bedding such as wood shavings.
Dirty, sweaty coat, shedding, boredom, poor blanket fit.	**Manageable.** Keep horse clean with regular grooming and use properly fitted blanket. If still persists, remove blanket or muzzle.
Greedy or was with competitive horses during feeding.	**Manageable.** Feed hay first. Put rocks or large feed wafers with feed in large shallow tub rather than deep bucket. Use large feed pellets or wafers instead of finely ground grain.
Unresolved stress, mimicry and then addiction. Theory: endorphins, which stimulate the pleasure center of the brain, are released during the behavior, leading to addiction.	**Manageable but incurable.** Means of dealing with it include cribbing strap that prevents contraction of neck muscles; also available with clamps, spikes, electric shock. Possible future pharmacological treatment. Surgery possible. Muzzle can be used in some situations.
Hormone imbalance, nasty disposition, insecurity, mare behavior.	**May be incurable,** as it is difficult to referee. Change turnout companion or may never be able to turn horse out with other horses.
Sexual frustration.	**Manageable.** Be sure horse has adequate exercise; can use mechanical devices to prevent erection.
Confinement, boredom, excess feed.	**Curable.** Provide exercise, diversion; don't use ground feeders and waterers; use rubber mats; don't reinforce by feeding.
Possibly endorphin addiction similar to cribbing; can be triggered by confinement, lack of exercise, or sexual frustration.	**Manageable/might be curable.** Geld nonbreeding stallions; increase exercise, reduce confinement; include stall companion or toy; neck cradle; muzzle; possible future pharmacological treatment. Professional management and training.
Confinement, impatience, likes to hear sound, doesn't like neighbor, gets attention.	**Can be curable,** depending on how long-standing the habit. Increase exercise, pad stall walls or hooves, don't reinforce by feeding, in desperation can try kicking chains or balls strapped to fetlocks.
Initially, dirty udder, sheath, or tail; shedding hindquarters; external parasites. Later, just habit.	**Manageable** with grooming, cleaning sheath and udder, worming, other medical treatments. For chronic habit, use electric fence. Wrap tail when trailering.
Confinement, boredom, excess feed, high-strung or stressed horse; can be socially contagious.	**Manageable.** Decrease grain, increase exercise. Turn out where he can be with or see other horses. Use specially designed stall door for weaver. Consider stall companion or toy.
Lack of course roughage in diet, boredom, teething, stress, habit.	**Manageable.** Increase roughage. Cover wood with antichew product or use metal or electric fence. Increase exercise, activity, and pasture time.

Bad Habits at a Glance

BAD HABIT	DESCRIPTION
Balking	Refusing to go forward often followed by violent temper if rider insists.
Barn-sour/ herd bound	Balking, rearing, swinging around, screaming, and then rushing back to the barn or herd or companion.
Biting	Nibbling with lips or grabbing with teeth, especially in young horses.
Bolting when turned loose	Wheeling away suddenly before halter is fully removed.
Bucking	Arching the back, lowering the head, kicking with hind legs, or leaping.
Can't catch	Avoiding humans with halter and lead.
Can't handle feet	Swaying, leaning, rearing, jerking foot away, kicking, striking.
Crowding	Pushing into the handler in the stall or while being led.
Halter pulling	Rearing or setting back when tied, often until something breaks or horse falls and/or hangs by halter.
Head-shy	Moving head away during grooming, bridling, clipping, vet work.
Jigging	Walking/jogging with short, stilted stride, hollow back, and high head.

CAUSES	TREATMENT
Fear, heavy hands, stubbornness, extreme fatigue, overworked.	**Curable.** Review forward work with in-hand and longeing, but don't overwork. Turn horse's head to untrack left or right. Strong driving aids with no conflicting restraining aids (no pull on bit). Do not try to force horse forward by pulling; you'll lose.
Separation from buddies or barn (food, comfort).	**Curable,** but stubborn cases require a professional. A confident, capable trainer gets the horse to progressively go farther from the barn (herd) and then positively reinforces the horse's good behavior, so horse develops confidence. The lessons *Go* and *Whoa* must both be reviewed.
Greed (treats), playfulness (curiosity), or resentment (irritated or sore). Investigating objects with mouth. Often from hand feeding treats or petting on the nose.	**Curable.** Handle lips, muzzle, and nostrils regularly in a businesslike way; when horse nips, tug forcefully on halter, then resume activities as if nothing happened.
Poor handling; anxious to exercise or join other horses.	**Curable but dangerous,** as horse often kicks as he wheels away. Use treats on ground before you remove halter; use rope around the neck.
High spirits, get rid of rider or tack, sensitive or sore back, reaction to rider's legs or spurs.	**Generally curable.** Monitor feed and exercise; use proper progressive training; check tack fit.
Fear, resentment, disrespect, bad habit.	**Curable.** Take time to properly train, use walk-down method in small area first, progress to larger. Remove other horses from pasture, use treats on ground; never punish horse once caught.
Insufficient or improper training. Horse hasn't learned to cooperate, balance on three legs, take pressure and movement of farrier work.	**Curable,** but persistent cases require a professional. Thorough, systematic conditioning and restraint lessons: pick up foot, hold in both flexed and extended positions for several minutes while cleaning, grooming, rubbing leg, coronary band, bulbs, and so on.
Poor training and manners.	**Curable** with proper in-hand lessons about personal space.
Rushed, poor halter training; using weak equipment or unsafe facilities, so horse gets free by breaking something. Often horse was tied by bridle reins.	**Can be curable but very dangerous,** and incurable in some chronic cases, which require a professional. Use long rope through tie ring and hold end, or use specialized tie ring; run rope around throatlatch and tie with bowline; use wither rope.
Initially, rough handling or insufficient conditioning; painful ears or mouth problems.	**Curable.** First eliminate medical reasons, such as ear, tongue, lip, or dental problems. Start from square one with handling; after horse allows touching, then teach him to put head down.
Poor training attempt at collection, horse not trained to aids, too strong bridle aids, sore back.	**Curable.** Check tack fit, use aids properly, including use of pressure/release (half halt) to bring horse to walk, or use driving aids to push horse into active trot.

Bad Habits at a Glance *(continued)*

BAD HABIT	DESCRIPTION
Kicking	Lashing back at a person with one or both hind legs; also "cow kicking," which is lashing out to the side and front.
Rearing	Standing on hind legs when being led or ridden, sometimes falling over backward.
Refusal to load	Balking, rearing, or backing up when asked to step into a trailer.
Running away/ bolting	Galloping out of control.
Shying/spooking	Spooking at real or imagined sights, sounds, smells, or occurrences.
Striking	Taking a swipe at a person with a front leg.
Stumbling	Losing balance or catching the toe on the ground and missing a beat or falling.
Tail wringing	Switching and/or rotating tail in an irritated or angry fashion.

USES	TREATMENT
...ally, reflex to touching legs, then fear ...fense) of rough handling or to get rid of ...reat or unwanted nuisance.	**Curable,** natural reflex that is easily overridden with progressive handling. Serious cases are very dangerous and require a professional to use remedial restraint methods.
...r, rough handling, doesn't think he must go ...ward or is afraid to go forward into contact ...n bit; associated with balking; a response to ...ected work.	**Can be curable** but is a very dangerous habit that might be impossible to cure even by a professional. Check to be sure no mouth or back problems. Review going forward in hand and review longeing.
...r training.	**Curable** with progressive lessons in leading, restraint, *Whoa,* and *Walk on.*
...r, panic, (flight response), lack of training ...he aids, overfeeding, underexercising, pain ...n poor-fitting tack.	**Might be curable but very dangerous,** as when horse panics, can run into traffic, over cliff, through fence, and so on. Remedy is to pull (with pressure and release on one rein) the horse into a large circle, gradually decreasing the size (doubling or one-rein stop).
...r (of object or of trainer's reaction to horse's ...avior), poor vision, head being forcibly held ...horse can't see, playful habit.	**Generally curable.** Put horse on aids and guide and control his movement with driving and restraining aids; thorough sacking out.
...ction to clipping, first use of chain or twitch, ...traint of head, dental work.	**Curable but very dangerous,** especially if coupled with rearing, as person's head could be struck. Hobbling by a professional.
...akness, lack of coordination, lack of condi-...g, young, lazy, long toe/low heel, delayed ...ak-over of hooves, horse ridden on forehand, ...r footing.	**Curable.** Have hoof balance assessed; check break-over; ride horse with more weight on the hindquarters (collect), conditioning horse properly.
...e back from poor-fitting tack, poorly ...anced rider, injury, rushed training, mare ...odiness, habit.	**May not be curable once established.** Proper saddle fit; rider lessons; massage and other medical therapy; proper warm-up; and progressive, achievable training demands.

Horse Timelines

K NOWING HOW A HORSE MATURES physically and mentally, and how it compares to our human maturity timeline, can help you design appropriate management and training programs.

Today's domestic horse experiences rapid physical maturity in the first two years, then levels off until the senior years, after which the decline can be quick unless the management is excellent. Exactly when the senior years occur is variable and depends on a horse's genetics and his care throughout his life.

Although some parallels can be drawn between characteristics of developing children and foals, at peak mental capacity a horse doesn't have even the intellectual capacity of a human toddler. The mental equivalent comments that follow are intended to show how a horse's thinking patterns develop as he matures.

Life-Stage Characteristics

Throughout a horse's development, he will display specific behaviors and physical characteristics indicative of his age. Knowing what is normal or average for a particular age will give you a better idea of what a horse needs and what type of behavior he is likely to exhibit.

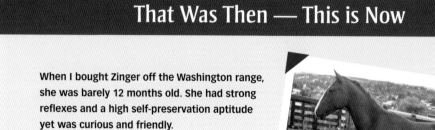

That Was Then — This is Now

When I bought Zinger off the Washington range, she was barely 12 months old. She had strong reflexes and a high self-preservation aptitude yet was curious and friendly.

During her prime, Zinger proved to be a powerful, trustworthy, and versatile horse. She was always there for me as a western, trail, ranch or dressage horse and as a broodmare.

At 31, senior Zinger is sound, willing, and wise. Although I'd like to keep riding her forever, I want her to enjoy her retirement in good health, so this photo was taken on our last ride.

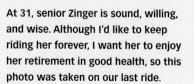

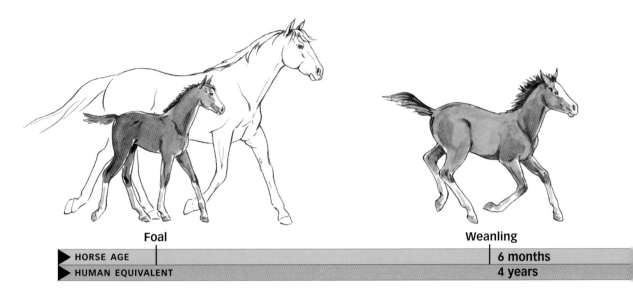

Foal

Weanling

| HORSE AGE | | 6 months |
| HUMAN EQUIVALENT | | 4 years |

Physical Growth

The physical development of horses progresses differently from that of humans. The timeline above, based on my experience and observation, is meant to serve as an approximation. For a rough idea of how a horse's development compares to a human's, one horse year equals eight human years until the horse is two years old; thereafter, each horse year equals two-and-a-half human years.

Foal

The foal is born with needs equivalent to a human infant's: he is preoccupied with hunger, thirst, sleep, and comfort. Being a precocial creature, however, within hours of birth the foal has the physical ability and mechanical skills of a two-year-old human. Twenty-four hours after birth the foal is able to run, using legs that are 90 percent as long as an adult horse's. Coupled with keen instincts, this physical advantage has helped the young horse survive over the millennia. Sometimes this physical development is expressed too exuberantly and foals stress themselves, especially when they are turned out following extended confinement.

In spite of their apparent vigor, foals are fragile, both mentally and physically, and need close contact and security from the dam. The suckling foal is characteristically inquisitive yet timid, fractious yet vulnerable, feisty yet fearful.

Handling the youngster from birth through the suckling period is beneficial, but make the sessions short, firm, fair, and to the point. The mare provides much of the foal's needed nutrition and immunity, but it is important to develop a conscientious feeding and health plan tailored to the foal.

Weanling

By weaning time at four to six months of age, the horse has reached the human physical equivalent of a four-to-five-year-old child and the mental equivalent of a two-to-three-year-old. With a short attention span and unpredictable outbursts, weanlings are best left to be horses, with lots of turnout time.

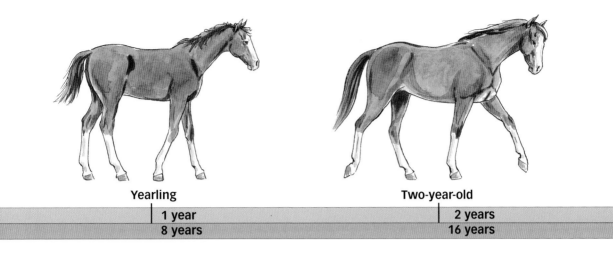

Yearling	Two-year-old
1 year	2 years
8 years	16 years

Lessons should be frequent, short, safe, and fun. The weanling can experience deep mental and physical trauma and is very impressionable, so harsh handling is never appropriate. Care must be taken to preserve the weanling's interest in eating and other routines so that he does not become depressed. The young horse separated from his mother is uncertain about his safety. In addition, he must rely totally on his own behavior patterns for the first time. If he is out on pasture, he must decide when and where to graze, drink, and find salt. If he is in a pen or stall, he must approach a new feeder or water trough on his own, without reassurance from his dam. Appropriate nutrition and health care are essential during this peak growing period.

Yearling

The yearling spends much of his time experimenting with his physical capabilities and finding his place in equine and human societies. The equivalent mentally of a five-year-old human and physically of an eight-year-old, the yearling horse can be testy, rambunctious, and moody. Fillies and colts are beginning to experience the effects of the hormones of puberty, so they add sexually oriented games into playtime. This is the age when many male horses are gelded.

It is imperative that the lessons started as a foal be reviewed and continued with the yearling. Although sessions are still short, they can be more frequent and cover a wider variety of handling. The yearling is receptive and capable of learning all of the domestic horse ground rules. He has nutritional

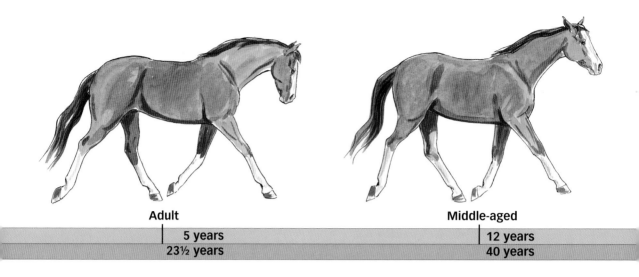

Adult	Middle-aged
5 years	12 years
23½ years	40 years

and health needs unique to his age, so he shouldn't be lumped in with adults' herd management just yet. For example, the yearling will require higher protein feed and may need to be dewormed more often than adults.

Two-Year-Old

With the two-year-old comes a serious sex drive and its subsequent effect of surging hormones on a colt's or filly's attention during training. Geldings tend to be more stable.

The physical equivalent of a sixteen-year-old human and the mental equivalent of a twelve-year-old, the two-year-old horse is too often treated as a mature horse. Many of the epiphyseal closures in the two-year-old's leg joints have matured, but he should not be made to accept the workload of an adult; his skeletal immaturity leaves him prone to injury. He lacks the stamina and strength to perform under a rigorous schedule without risking permanent physical or mental damage.

Having lost much of the yearling's silliness, the two-year-old can generally focus on lessons and show the trainer his potential. Until the age of five, the horse's teeth are shedding and erupting, and his epiphyseal closures continue to mature.

Adult/Prime

The five-to-twelve-year-old horse is in his prime both physically and mentally. The physical and mental equivalent of a human in his twenties and thirties, the adult horse has a mature physique and has had many experiences that (hopefully) have made him sensible. His nutritional requirements

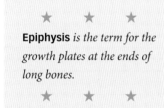

★ ★ ★

Epiphysis *is the term for the growth plates at the ends of long bones.*

★ ★ ★

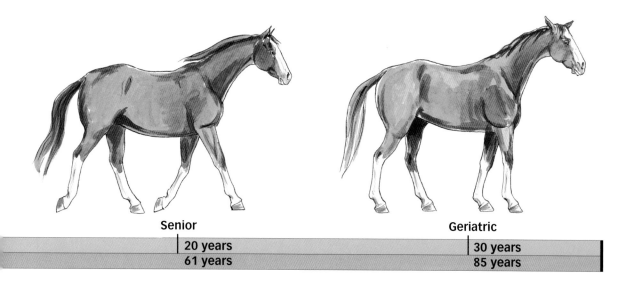

Senior	Geriatric
20 years	30 years
61 years	85 years

are basic. He can remain physically fit with moderate exercise, and his immune system is at its peak.

Middle-Aged

Between prime and senior years, the twelve-to-twenty-year-old horse shows noted changes in stamina and can lose some muscle tone. Similar to forty-to-sixty-year-old humans, the changes are extremely variable between horses, depending on their genetics, use, and care.

Senior

The senior horse, at age twenty to thirty, like his over-sixty human counterpart, will likely have dental changes (lost, worn, and/or broken teeth) that lead to difficulties in chewing, assimilating nutrients, and maintaining weight. Dietary needs include an increase in the amount of feed and the quality and amount of protein and fats, but a decrease in carbohydrates. Vision and hearing deterioration can affect behavior. The immune system is not as strong as it once was, and arthritic changes may cause lameness.

Geriatric

Horses of more than thirty years of age might be physically impaired, but many, like Zinger, remain healthy enough to enjoy light work, driving, or riding. The biggest challenge is feeding, as they often require easily chewed feeds such as mash, beet pulp, and hay chop. Regular low-level exercise is essential to prevent stiffness just as for humans over eighty-five.

Mental Equivalent

Since a horse never even approaches the intellectual capacity of a human, when I use the phrase "mental equivalent" I am referring to a parallel stage or a comparable phase between horses and humans. As an example, a weanling has mental characteristics similar to that of a 2–3-year-old child — a short attention span and sudden outbursts. That's what I mean by mental equivalent.

The Timeline of a Horse's Life

Birth

Has the mechanical skills (legs) for survival; stands and nurses at birth; weighs approximately 10 percent of mature weight yet legs are 90 percent of mature length and height is 75 percent of mature height; inquisitive; playful; sleeps a lot (see Newborn Foal Timeline)

weight 110 pounds
height 11.2 hands

Human Equivalents:
Physical – 2 years
Mental – Newborn

4 months

Great rate of physical growth; playful and tests physical limits; may start sex behavior with dam; gains confidence until weaning, then insecure; chews objects

weight 400 pounds
height 13 hands

Human Equivalents:
Physical – 4 years
Mental – 2 years

6 months

Great rate of physical growth; about half of adult weight; impressionable, curious, reactive; can be insecure or feisty, frisky; begins sex play in earnest; short attention span, chews objects

weight 500 pounds
height 14 hands

Human Equivalents:
Physical – 5 years
Mental – 3 years

10 years

Prime of life; usually at 12 years of age, horse is called "smooth-mouthed" (see pp. 101–102)

Human Equivalents:
Physical – 36 years
Mental – 40 years

6 years

Beginning of physical and mental prime, which lasts to age 12 to 15, depending on horse

Human Equivalents:
Physical – 30 years
Mental – 30 years

15 years

Mental aging starts slowing down while physical aging continues at same rate; might be a little past prime or at beginning of middle age; a little stiffness; an additional sensitivity to weather and insects; becoming more tolerant and relaxed about humans and use

Human Equivalents:
Physical – 48½ years
Mental – 45 years

20 years

Bones become more brittle, wear and tear shows on joints in beginning stages of arthritis; vision deteriorates; teeth stop erupting and wear begins approaching the gum line; begin graying around eyes, ears, and muzzle; mare fertility declines

Human Equivalents:
Physical – 61 years
Mental – 50 years

25 years

The average lifespan of a horse; may show signs of swayback and hay belly, decreased muscle tone and weight, prominent backbone; lose teeth; drooping lower lip; dry skin; less saliva production and ability to absorb nutrients; may be anemic; winter coat comes in early, grows long, and sheds later

Human Equivalents:
Physical – 73½ years
Mental – 55 years

1 year

Sexual display; gelding often occurs; testy, moody; chewing; biting; wolf teeth often removed

weight 600 pounds
height 14.2 hands

Human Equivalents:
Physical – 8 years
Mental – 5 years

18 months

Sexually mature; estrous occurs in fillies, and males capable of breeding

weight 875 pounds
height 15 hands

Human Equivalents:
Physical – 12 years
Mental – 8 years

2 years

Serious sex drive; growth plates may have matured enough for light work; light riding is often started

weight 925 pounds
height 15.2 hands

Human Equivalents:
Physical – 16 years
Mental – 12 years

5 years

Considered first year as a mature horse: has full mouth of permanent teeth (see pp. 101–102); mature skeleton (see p. 99); many 5-year-olds are already solid adults

Human Equivalents:
Physical – 23½ years
Mental – 25 years

4 years

From this point on, weight and height are usually stable until senior years

weight 1100 pounds
height 15.2 hands

Human Equivalents:
Physical – 21 years
Mental – 21 years

3 years

Will vary from still a bit silly to settled and developing adult habits

weight 1050 pounds
height 15.2 hands

Human Equivalents:
Physical – 18½ years
Mental – 18 years

30 years

Joints relax, so pasterns slope more; eyes may become cloudy; partial blindness

Human Equivalents:
Physical – 86 years
Mental – 60 years

35 years

May have lost a number of teeth by now, so needs soft food

Human Equivalents:
Physical – 98½ years
Mental – 65 years

40 years

May be stiff and have difficulty eating, but is a treasure

Human Equivalents:
Physical – 111 years
Mental – 70 years

A new foal is a clean slate onto which you can write a masterpiece.

Development Timelines

Knowing *what* happens *when* with your horse can help you in many ways: to design a training and health program, to choose tack, to conduct an optimal exercise and training program, and much more.

Newborn Foal Timeline

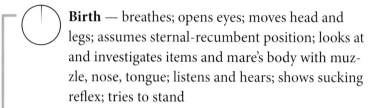

Birth — breathes; opens eyes; moves head and legs; assumes sternal-recumbent position; looks at and investigates items and mare's body with muzzle, nose, tongue; listens and hears; shows sucking reflex; tries to stand

1 hour — stands; seeks dam; walks; defecates meconium (a newborn's first stool); follows dam; nurses; nickers; imprints to dam; resists restraint; shows withdrawal reflex

2 hours — lies down (sternal-recumbent and lateral-recumbent positions); sleeps; gets up; fears new objects or people but is curious, and secure with dam, so investigates

3 hours — begins to play; self-grooms sides; moves tail; tests objects with mouth; trots; gallops

24 hours — scratches head with hind hoof; rubs on objects; yawns; makes clacking (snapping) mouth movement; rolls; exhibits flehmen response

Growth-Plate Closures

Here are the ranges of growth-plate closure times in the equine forelimb. **Distal** refers to the lower growth plate of a particular bone. **Proximal** refers to the upper growth plate of a particular bone.

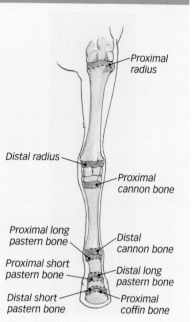

Birth	Proximal coffin bone Distal short pastern bone Distal long pastern bone Proximal cannon bone
6–15 months	Proximal short pastern bone Proximal long pastern bone
6–18 months	Distal cannon bone
11–25 months	Proximal radius (forearm)
22–42 months	Distal radius
26–42 months	Proximal humerus (arm) (not shown)
36+ months	Proximal scapula (shoulder blade) (not shown)

Proximal radius

Distal radius

Proximal cannon bone

Proximal long pastern bone

Distal cannon bone

Proximal short pastern bone

Distal long pastern bone

Distal short pastern bone

Proximal coffin bone

Joint Closure Timeline

The horse's skeleton is not fully mature until between the fourth and fifth years. It is critical to know whether joint closure has occurred at the distal radius (see illustration above) before he begins hard work. If you want to start working a horse before two years of age, have his knees x-rayed and the radiographs interpreted by your veterinarian. If work begins too early, the young horse can suffer from epiphysitis, or inflammation of the epiphyseal closures. Such a condition can predispose a horse to lameness and limb deformities.

Dental Timeline

All mature horses have incisors, premolars, and molars. Some also have wolf teeth and canines, which will be discussed separately. An adult horse will have at least thirty-six teeth. There are twelve incisors at the front of the mouth, twelve premolars beginning at the corner of the lips, and twelve molars at the back of the mouth.

Before a horse develops his permanent teeth, he has a set of temporary teeth. For example, the temporary premolars are in place by two weeks of age but are replaced by the permanent

A young (2-year-old) horse's permanent incisors and molars are beginning to replace the temporary ones.

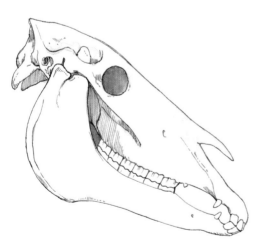

An aged (5-year-old or older) horse has a complete set of incisors and molars, and a male will most likely have canine teeth as well.

premolars between the ages of two and five years. Most age guidelines are based on time of tooth eruption. Teeth begin to wear six months after they erupt.

Premolar does not refer to a temporary tooth but to the position of the tooth in the mouth; premolars are located ahead of the molars. When a particular tooth appears, such as the second premolar, all four second premolars erupt at approximately the same time: the upper right, the upper left, the lower right, and the lower left.

Between the premolars and the incisors is a relatively toothless space called the interdental space, which is where the bit lies. After the age of four, however, most male horses also have four canine teeth (tushes) in the interdental space behind the incisors. Male canines usually begin to erupt at four years of age and are most often fully developed at five. Canines can get very sharp, so they need to be clipped or rasped to keep them from cutting the horse's lips when bridled. Very few mares develop tiny canine buds.

In addition, some horses have wolf teeth (first premolars) in front of the second premolars, but usually only in the upper jaw. Through evolutionary processes, wolf teeth are absent in some horses. If they are going to appear, they will erupt by the time he is a yearling. Although they are permanent teeth, sometimes wolf teeth are pushed out (along with the temporary second premolar) when the permanent second premolar erupts at two to three years of age. They can sometimes cause painful pinching of the lip skin when a snaffle bit is used on the horse. Because of this, veterinarians often extract wolf teeth.

The numbering of temporary and permanent premolars and molars can be confusing, since the wolf tooth is called the first permanent premolar. Refer to the chart on pages 102–103 for clarification.

Until a horse is five years old, his teeth are constantly erupting, shedding, and being replaced. At five he is said to be aged and to have a full set of teeth. His teeth will continue to emerge, however, until he is in his early twenties.

By the time a horse is five years old, in addition to the teeth you can see above the gum line, below each tooth there are 3½–4 inches of extra tooth embedded in the bone of the jaw. These are called reserve crowns. As a horse ages, these reserves emerge continuously at about the same rate that the surface wears away from chewing and grinding. By the time a horse is in his late twenties, he starts running out of reserve crowns, and by his late thirties he could have lost many of his teeth and be chewing mostly with gums.

Wear and Tear

The upper jaw of a horse is 30 percent wider than the lower jaw. As the horse grinds his food with a sideways motion, he wears down his molars, causing sharp edges to form on the outside (cheek surfaces) of the upper molars and premolars, and on the inside (tongue surfaces) of his lower molars and premolars. These sharp points can interfere with a horse's chewing and can cause cheek and tongue lacerations. That's why it is important to schedule an annual dental exam for each horse so your veterinarian can file off the points (a process called floating) and keep the wear on the teeth balanced.

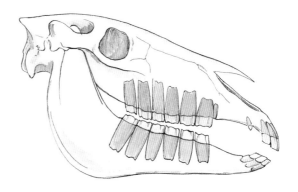

The reserve crowns are the parts of the teeth that lie below the gum line. These grow during a horse's life but can be used up by the time he reaches old age.

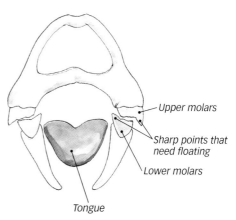

Upper molars

Sharp points that need floating

Lower molars

Tongue

A cross section of the horse's jaw from the front, showing the sharp points that form from normal chewing.

Tooth Development

AGE	TOTAL # TEETH	INCISORS temporary	permanent
Birth to 2 weeks	16	Centrals appear at birth or erupt by 2 weeks; 4 total	0 total
4–6 weeks	20	Intermediates erupt; 8 total	0 total
6–9 months	24	Corners erupt; 12 total	0 total
1 year	28	12 total	0 total
2 years	32	12 total	0 total
2½–3 years	32	Centrals pushed out; 8 total	Centrals erupt; 4 total
3½–4 years	36	Intermediates pushed out; 4 total	Intermediates erupt; 8 total
4½–5 years	36	Corners pushed out; 0 total	Corners erupt; 12 total
7 years	36	0 total	7-year hook on upper corner incisors; 12 total
9–10 years	36	0 total	Galvayne's Groove appears at gumline of upper corner incisors; 12 total
12 years	36	0 total	Galvayne's Groove extends ¼ way down upper corner incisors; 12 total
15 years	36	0 total	Galvayne's Groove extends ½ way down upper corner incisors; 12 total
20 years	36	0 total	Galvayne's Groove extends entire length of upper corner incisors; 12 total
25 years	36	0 total	Galvayne's Groove ½ gone; 12 total
30 years	36	0 total	No Galvayne's Groove; 12 total

| PREMOLARS | | MOLARS | |
temporary	permanent	permanent	COMMENTS
#2–4 on each side, top and bottom; 12 total	0 total	0 total	
12 total	0 total	0 total	
12 total	Variable/ Wolf Teeth; 0–4 total	0 total	Wolf teeth (1st Permanent Premolar) may appear in front of the 2nd Temporary Premolar usually of the upper jaw
12 total	Wolf Teeth; 0–4 total	#1 erupts; 4 total	Wolf teeth often removed
12 total	0 total	#2 erupts; 8 total	
# 2 & #3 pushed out; 4 total	#2 & #3 erupt; 8 total	8 total	
#4 pushed out; 0 total	#4 erupts; 12 total	#3 erupts; 12 total	Sometimes temporary molar caps need to be popped off by vet
0 total	12 total	12 total	0–4 Canines appear in the interdental space in males and some mares; horse has "full mouth"
0 total	12 total	12 total	
0 total	12 total	12 total	
0 total	12 total	12 total	Horse is called "smooth mouthed" which means all the cups on the chewing surfaces of the incisors are gone
0 total	12 total	12 total	
0 total	12 total	12 total	May begin losing teeth; incisors appear more angled forward when viewed from the side
0 total	12 total	12 total	Teeth might be worn to gumline
0 total	12 total	12 total	Missing and worn teeth require special feed

Communication

SINCE HORSES COMMUNICATE primarily through body language and feel, they interpret our actions in their own terms. That's why we need to learn their language — not only to understand what horses are saying to us, but also to know what our position, posture, and movements are saying to them.

Each time you interact with your horse, the two of you are having a conversation. Just as with people, conversations can go well or badly. Here are two versions of a typical conversation between a horse and his trainer. Which one do you think will have the best long-term results?

HEAD DOWN

TRAINER: Reaches for forelock. ("Let me straighten this out for you, Star.")

HORSE: *Raises head quickly and very high. ("There's something coming at my head; I have to protect my eyes and ears.")*

TRAINER: Can no longer reach forelock, so jerks down forcefully and repeatedly on lead rope, which causes pain to Star's nose and poll from the halter. ("Do what I want! Don't pull away from me!")

HORSE: *Raises head higher. ("I didn't move fast enough, so now there's pain on my nose and head. I'm really scared.")*

TRAINER: Continues jerking, begins yelling. ("By golly, you are going to lower your head, you stupid horse.")

HORSE: *Starts swinging his head off to one side and the other. ("I'll try something else to get rid of this pain on my head and nose. What's going on?")*

TRAINER: Slaps the horse on the neck with the end of the lead rope. ("I'll show you who's boss. See what happens when you don't do what I want!")

HORSE: *Swings his head over to the right and rears 2 feet off the ground. ("Swinging my head didn't work; maybe if I go higher I can get away from the pressure.")*

TRAINER: Jerks on rope but loses grip. ("I'll teach you to disobey me.")

HORSE: *Pulls away and trots 30 feet away, turns and stops, looks at trainer with his head down. ("Well, that worked, all I have to do next time is rear and pull away to stop the pain. Glad I finally figured that out.")*

TRAINER: Reaches for forelock. ("Let me straighten this out for you, Star.")

HORSE: *Raises head quickly and very high. ("There's something coming at my head; I have to protect my eyes and ears.")*

TRAINER: Can't reach forelock, so applies steady downward pressure on the lead rope to the halter to the horse's poll. ("Put your head down a little, Star. I can't reach your forelock.")

HORSE: *Raises head 1 inch, then lowers head half an inch. ("I don't like that pressure on my head. I'll try moving my head up to get away from it. That didn't work, so I'll try lowering my head just a little bit.")*

TRAINER: Holds steady pressure when head rises, then instantly releases pressure when head begins to lower, and rubs Star on the forehead. ("Good boy.")

HORSE: *Keeps head still. ("This feels good.")*

TRAINER: Applies downward pressure. ("It's too hard to reach your forehead with your head way up there, Star. Bring your head down a little more.")

HORSE: *Lowers head 1 inch. ("Last time I felt pressure, it went away when I lowered my head.")*

TRAINER: Instantly releases pressure and rubs Star on the forehead, saying "Good boy." ("Now you're learning. What a smart horse!")

HORSE: *Lowers his head 2 more inches. ("When I lower my head it feels good.")*

TRAINER: Moves hand from forehead to forelock. ("Don't worry, Star, it's just like the rubbing.")

HORSE: *Tenses muscles in neck and starts raising head. ("I'm not sure about that thing coming at my ears, but I haven't been hurt yet.")*

TRAINER: Applies light pressure instantly, keeping hand on forelock. ("Figure it out, Star.")

HORSE: *Lowers head 5 inches. ("If I lower my head, the pressure goes away. It's worked every time so far. And I haven't been hurt.")*

TRAINER: Releases pressure when Star lowers head. Continues rubbing forehead and forelock while saying "Good boy." ("That's enough for right now. You did really well.")

BENDING

RIDER: With snaffle rein in hand, moves arm out to the right, which puts tension on the rein. ("Bend your head to the right, Star.")

HORSE: *Feels pressure on the corners of the lips on the right side of his face and from the bit ring on the left side of his face. ("I don't like this pressure; what can I do to avoid it?")*

Begins to move his head and neck to the right. ("I wonder if the pressure would stop if I turned my head this way.")

RIDER: Takes up the slack and maintains pressure by pulling his hand back and out to maintain a constant pressure on the horse's mouth. ("It's working; Star bent his head to the right. Now let's get a little more.")

HORSE: *Bends more to the right, then tries straightening his head and neck, but hits the bit so raises his head. ("Well, bending to the right doesn't make the pressure stop. Maybe if I go the other way. No, that's worse. Maybe if I raise my head.")*

RIDER: Maintains pressure and adds a downward pull to the rein pressure. ("C'mon, Star, don't be stupid. You had it there a minute ago. Why did you start fighting? Maybe if I give you a stronger cue, you'll figure it out.")

RIDER: With snaffle rein in hand, moves arm out to the right, which puts tension on the rein. ("Bend your head to the right, Star.")

HORSE: *Feels pressure on the corners of the lips on the right side of his face and from the bit ring against the left side of his face. ("I don't like this pressure; what can I do to avoid it?")*

RIDER: Holds light contact. ("C'mon, Star, figure it out.")

HORSE: *Looks to the right, tips right ear to the right. ("I wonder if the pressure would stop if I turned my head this way.")*

RIDER: Maintains light contact but does not add more. ("I can see the wheels turning, Star. You can do it.")

HORSE: *Begins to move his head and neck to the right. ("I'll move this way and see if the pressure goes away.")*

RIDER: Feels the first instance of compliance in the neck, jaw, and poll and adjusts (yields) the pressure appropriately on the rein as soon as the horse starts to soften. ("I felt Star give, so I have to reward immediately if I want him to make the connection.")

HORSE: *Momentarily holds the bend or bends a little more to the right. ("The pressure went almost completely away when I bent my head and neck this way, so it's OK here.")*

Reading a Horse's Body Language

You can tell what a horse is thinking by looking at his overall stance, the position of his head and neck, the use of his ears, and the action of his tail and legs. In general, you will be able to tell if the horse is fearful, passive, assertive, or aggressive. You will know whether he is welcoming you or telling you to stay out of his space. Once you know the larger, more graphic signals, you will become aware of the subtler precursors to those movements. (See The Subtleties, page 121.)

Overall Stance

How can you tell if your horse is content, fearful, or agitated? His overall stance will give you a good starting point.

Content and Relaxed

A relaxed horse is often standing with one hind leg resting, his head and neck in a soft, lowered position, his eyes soft or partially closed, his ears relaxed and maybe slightly tipped to the side, and his muscles relaxed. A relaxed horse is content and feels secure.

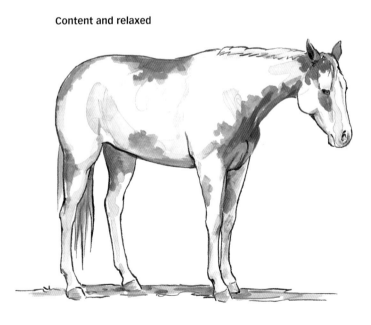

Content and relaxed

Friendly

Friendly

When a horse is friendly, his ears are forward. His head is at a medium height as he softly reaches with his muzzle, as if to sniff. His soft eyes look interested, and his muscles are relaxed.

Alert

An alert horse has his head up, his ears forward, his face line about 45 degrees in front of the vertical, and his nostrils open and actively taking in the scents in the air. The alert horse is often perfectly still and intent, but there is no fear in his eyes.

Unfriendly

It is hard to mistake an unfriendly horse. He holds his head low and reaches aggressively with his muzzle, perhaps with bared teeth and ears back. His eyes are cold and glaring, his nostrils pinched and wrinkled; perhaps his tail switches and one of his hind legs is raised in warning.

Alert

Unfriendly

Fearful

A fearful horse has his legs in a ready-to-flee position, or they are already in motion. His head and neck are up and alert, his eyes are wide open and his ears are facing the perceived danger, his muscles are tense and ready for instant flight. An anxious horse is worried and dangerous.

Balking

A horse that is balking is rigid and unyielding. He won't move, no matter what you do. He is mentally and physically tuned out. A horse usually balks because he is in pain, he is afraid, he is confused, or he has developed a bad habit.

Fearful

Balking

Bucking

A bucking horses alternates between lowering his head and standing on his forelegs, and raising his head and standing on his hind legs, often in a sunfish (rocking) motion either in place or as he moves forward. A horse bucks because he is afraid, the tack or rider is uncomfortable, because it feels good, or out of habit.

Hyper

The exuberance and high spirits of an overfed and underexercised horse can resemble that of a fearful horse. However, an experienced observer can usually perceive playfulness and confidence in the horse's facial expression — a twinkle in his eye rather than a wary eye.

Many behaviors are rider-induced. This horse, kicking out when the leg or whip is applied, is essentially saying: "I am not familiar with the feel of the whip and I don't like it. I don't know what it means. I feel like I am being asked to go and stop at the same time. I am confused."

Bucking

Hyper

Rearing

A rearing horse stands on his hind legs, sometimes so straight up that he could fall over backward, especially if he has the added weight of a rider on his back or if the rider is pulling back on the reins. A horse rears because he can't go forward or backward or doesn't want to. If he is restricted by tack or the rider and is given conflicting signals, maybe up is the only way he thinks he *can* go. On the other hand, he might use rearing as a power move to make himself bigger and say no.

Sick

If you know your horse well and he suddenly acts differently than normal, you will suspect that something could be wrong physically. A sick horse tends to either lie or stand very still, or to move around violently, pawing, rolling, and looking at his flank. Pain might make him extremely passive and non-reactive, or he might buck, rear, avoid being tacked up, react violently to cinch tightening, avoid being mounted, refuse to move forward, stumble, put his ears back and dive at you, or work with a hollow back, raised head, and short stride.

Bolting

Horses that run suddenly and fast and are hard to stop are bolting. They often bolt in conjunction with a spook. Bolting is most commonly caused by fear or extreme insecurity and disobedience. Horses that are herd bound or barn-sour often bolt to return to the herd or barn.

Rearing

In Context

It would be useless to write a dictionary of absolute definitions of horse body language because, for example, "ears back" can mean anything from ill to paying attention to angry. We have to read these signs in context, that is, along with what else the horse's body is doing.

For instance, one day, my vet, who knows my horses well, came to float their teeth. I noticed his new assistant look with concern at Zinger who has one eye with more white than normal. Usually when a horse shows the whites of his eyes, it means he is excited or fearful but Zinger was standing relaxed on 3 legs with her head down. So, in context, the white of her eye was not a concern.

Bolting

Sick

Position of Hindquarters

Since a horse's hind legs are a means of protection from predators, the position and activity of the hindquarters tell you a lot about what he is feeling. Here are some insights.

Hindquarters resting

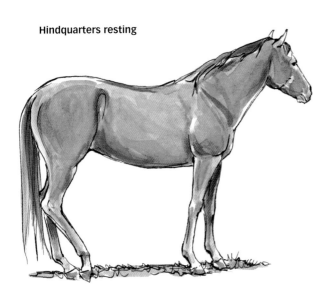

★ When a horse is resting, his croup is lowered, often with a resting hind leg, and the hindquarter muscles are soft.

★ When a horse relates to another horse or a human, the orientation of his hindquarters indicates their relationship. When a horse faces you, he is inviting. When a horse presents his hindquarters to a human or another horse, it is generally an act of protection or aggression, as it often precedes a kick.

★ When a horse is standing in a storm, he turns his hindquarters toward the weather and lowers his head. The body mass of his thickly muscled hindquarters provides a wind and rain/snow block for his vulnerable head.

Threatening

One exception to usual hindquarter language is when a mare wants to be bred by a stallion: she will present her hindquarters to him for breeding. Even though she might be soliciting the breeding, however, she might still kick the stallion. Another is when a young foal has had his tail head scratched by humans (something foals dearly love), but turning his rear to humans can become a dangerous habit. If the foal turns his hindquarters to the human every time a human approaches, even if he is only asking for scratching, it can be a hard habit to break.

Head and Neck

The level and action of a horse's head and neck are often closely tied with the action of the hindquarters and indicate what is likely to be coming soon. Here's what to look for.

⭐ A level, relaxed head and neck mean that there is likely to be no action soon. The horse is content and hanging out.

⭐ When the head is elevated or in constant motion, going up and down or side to side, it is important to pay attention.

⭐ A particularly aggressive head gesture is when the head is held very low, often snaking back and forth with the ears pinned flat back and the teeth bared. You might see a mare exhibit this behavior when protecting her foal. It means, in no uncertain terms, "Stay away or I'll bite." The reason for the low head is so that the horse can protect her own legs and vital organs while biting the intruder's legs to disable him.

⭐ A sudden thrust forward of the head at any level means "Beware" or "Get out of my space."

⭐ On the other hand, when a horse slowly and smoothly reaches forward with his head, he is being inquisitive or curious or might be soliciting rubbing. When you groom your horse, he might raise and stretch his neck as if to say "Ah, that's the spot."

Alert

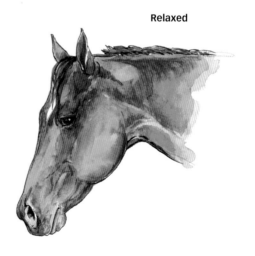

Relaxed

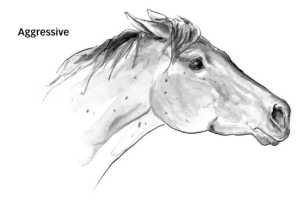

Aggressive

Bared

Clacking

Teeth

A horse's teeth are not to be trifled with. They are capable of removing a finger or a nose in an instant.

⭐ Bared teeth mean "Stay away; I'll bite."

⭐ Clacking of the teeth is a submissive gesture of a young horse toward the mouth of the dominant horse, similar to muzzle licking in dogs.

Lips

Although it is tempting to pet a horse on his soft lips, they are really for inspecting and eating. Here are some rules of thumb about lips.

⭐ When a horse's lips are closed but loose, the horse is relaxed. Licking and chewing indicate that a horse is relaxed and submissive.

⭐ When they are closed, pursed, and held tight, the horse is tense and probably not breathing effectively.

⭐ When a horse's lips are open, he is eating, drinking, yawning, examining an object, or getting ready to bite.

Nostrils

Next to the eyes and ears, the nostrils can tell you the most about how a horse is feeling.

⭐ Soft nostrils mean a horse is relaxed; flaccid could mean the horse is ill or bored.

⭐ Tight and rigid nostrils indicate tension from fear or pain or are a sign of aggressiveness.

⭐ When a horse's nostrils flare, he is either winded from exercising and trying to catch his breath, or they are taking in scents for processing.

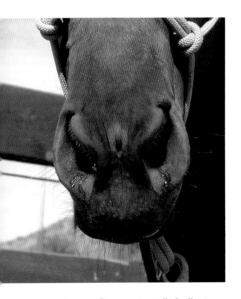

These soft, open nostrils indicate a horse that is calm and relaxed.

This mare is saying to the photographer, "Stay away from my foal or I'll go after you!" She also might be trying to prevent her foal from being too friendly with humans.

Off to the side (relaxed)

Ears

If you look at the ears in relation to the overall stance and the position of the head and neck, they can provide you with a great deal of information about the horse's temperament and attitude.

★ Usually when one or both ears are flopping off to the side, the horse is relaxed. This can occur when the horse is resting or when he is working.

★ Ears that are rigid can indicate a horse that is tense or alert, depending on the other body signs. When both ears are ahead and the horse is looking intently, that usually indicates alertness.

Flat back (resistance)

★ A horse with ears pinned flat back is showing anger, resistance, or aggression. A horse that is ill or in pain, however, can also pin his ears back. When a horse's ears are flat back, the ear canal is sealed off as a protective measure.

★ A horse that is being ridden might have one or both ears turned backward to his rider. This is often a sign of the horse respectfully paying attention to what is happening back there in one of his blind spots as well as keeping track of what is ahead.

One forward, one back
(paying attention)

Kind

Relaxed

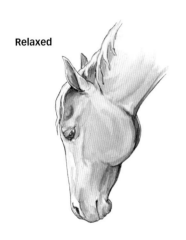

Frightened

Eyes

If there is one part of a horse that we are drawn to immediately, it is his warm, soft, large, dark eyes.

★ When a horse is content, secure, and relaxed, he has a soft, almost mystical look in his eyes of inward focus — part dreaminess, part deep sigh, and all very beautiful. This is how we hope to see our horses most of the time.

★ A horse that has been worked into the ground, has been treated unfairly, or is ill or injured can have a dull eye that almost looks sunken. He has retreated into himself and is tuned out. He is not listening, is listless, and barely responds. To bring a horse back from such a black hole is difficult, because he has given up.

★ Also unmistakable, however, is the hard, glaring eye that we might see in a horse that is in protective mode. A broodmare might have that look, for example, when she is protecting her foal. If your horse gives you this cold eye when you come to give him feed, however, he is trying to assert his dominance over you. You need to make it clear to him that you are the dominant one and you are also benevolent, and that is why you are providing feed. Horses should learn to accept feed submissively from humans.

★ A worried horse can sometimes show wrinkles around the eyes, as if concentrating on something such as minor pain or a small concern.

★ When a horse is frightened or in panic, he opens his eyes so wide that we sometimes see the whites of his eyes, the sclera. This often indicates he is ready to pop into action in an instant. Some horses, however, such as Appaloosas, have a smaller iris in relation to the sclera, so that even when relaxed, the white shows.

Leg Language

A means for locomotion, aggression, protection, and inspection, a horse's legs have a language of their own, which should be carefully read and taken seriously.

★ A softly raised and resting hind leg indicates relaxation.

★ A quickly raised leg indicates a threat: "Stay away or I'll kick."

★ Pawing indicates impatience or illness or is used to look for food or to prepare the ground for rolling.

★ Striking with a foreleg is a dangerous and aggressive behavior that indicates a horse feels threatened and wants to remove the other animals or object from his space.

★ When a horse stomps with any leg, it indicates impatience, anger, or irritation, such as with flies.

Raised, relaxed

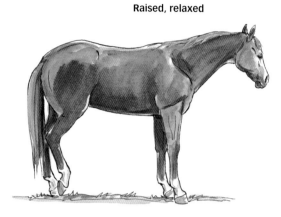

Raised to kick

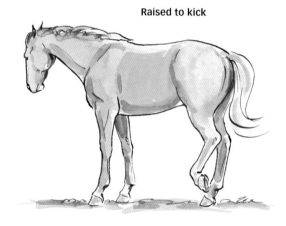

Striking

Tail Talk

The position and activity of the tail indicate the level of relaxation of the muscles of the hindquarters.

★ If the croup is rounded, "dropped," and relaxed, the horse's tail will be soft and swinging from side to side as he moves.

★ A clamped tail means tension, fear, and possible kicking or bolting to come.

★ If a horse's back is hollow and tense, his tail will be held in a tense, raised position.

★ A horse in high spirits might hold his tail straight up.

★ A raised tail can also indicate alertness.

★ A swishing (switching or flicking) tail indicates irritation. A flicking tail movement is characteristic of horses that don't like leg aids or spurs. A swishing tail can also be characteristic of a mare in heat telling other horses to stay away from her hind end or for a select few to come closer.

Swishing

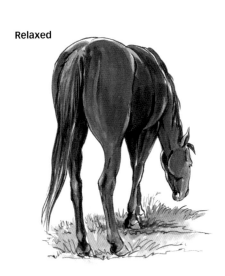

Relaxed

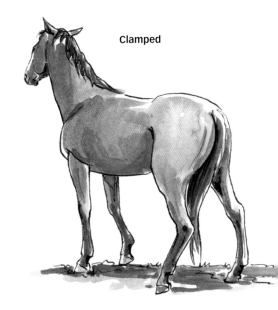

Clamped

The Subtleties

As you learn to read horses and become a keen observer of their body language, you'll find that often before a big movement, there are lots of little signals. If you see and understand the little things, you can progress faster and avoid unwanted escalation. If you reward when a horse is trying, it will encourage him. If you take a subtle step or shift your weight when you see a small warning sign, you just might head off a wreck.

Before a horse explodes, he will give you lots of little warning signs, such as:

looking away ★ tight jaw ★ ears pointing away from you ★ clamped tail ★ head up ★ leaning away ★ muscles contracting ★ turning away ★ lip tension ★ stepping away

Before a horse actually does what you are asking him to do, he may show you lots of little attempts or tries that you can reward, such as:

looking toward you ★ blowing ★ ears pointing toward you ★ relaxed tail ★ head down ★ leaning toward you ★ reaching toward you ★ turning toward you ★ loose lips ★ stepping toward you ★ licking lips

Emotion, Mood, Mental State, or Attitude?

You know your horse's temperament and his level of training and management, yet from day to day you see a difference in his body language and the way he behaves. These differences can be called a mood, an emotional or mental state, or, my favorite, an attitude, which is a temporary outlook that can be caused by a variety of factors.

How your horse greets you can be negatively influenced by illness, injury, pain, hormones, separation, fatigue, hunger, thirst, fear, or inclement weather. He can be positively influenced by good health, socialization, fitness, rest, satiation (with food and water), safety, confidence, and mild weather.

Just like you, your horse can have good days and bad days. You can minimize the number of bad days by being a conscientious horse keeper and trainer who fills her horse's needs and treats him fairly.

Vocal Language

Although body language is the main means of horse communication, horses express themselves vocally, as well. Here is a translation of some of these expressions.

Exhale. An outward sigh that is soft and relaxed means just what is sounds like: "Aaaaaahhhhh," a release of tension.

Sharp snort or blow. One or two snorts might be a punctuation of alarm, or they could just indicate that a horse is clearing dust from his nasal passages.

Vibrating, rolling snort. Usually uttered in deep tones, a rolling snort means a horse is very wary and suspicious and may suddenly bolt. This has always been the hallmark of Zinger, who lets us know when our behavior or something else is strange by her standards. It is more frequent now that her vision is failing a bit at age thirty-one.

Whinny or neigh. This is a loud call that usually starts high and drops in tone. It can often be heard as far away as half a mile, so it is something that your neighbors and their horses can hear. Horses use a full-volume whinny to make or maintain contact, as a warning, or to solicit attention or care. Of our seven current horses, five are relatively silent and two mares (half-sisters) call whenever there is a marker event, such as a door opening, the movement of another horse on the property, or merely the appearance of a person.

Scream. An intense calling is typical of a weanling who is trying to reestablish contact with his dam, or of a frantic, herd-bound horse.

Nickering. A soft, low chortling is how a mare greets her foal or a horse greets a human friend, such as at feeding time.

Grunt. When a horse groans or grunts, it is a sign of deep exertion. Some horses grunt when they roll or when they buck and kick.

Squeal. A short, high-pitched, excited call, often made by a mare in heat, can mean "Come here" or "Get away."

Blowing. Horses often say hello to each other by blowing into each other's nostrils. It may end almost as soon as it begins, or it may escalate to excited nickering, grunting, squealing, and varied displays of body language, friendly and otherwise.

At weaning, 4-month-old Sherlock calls shrilly to his dam Sassy who is several pastures away enjoying her freedom.

How to Communicate with Your Horse

Now that you know how horses talk, you can begin developing a means of communication that your horse will understand.

Body Language

Since horses begin reading your body language the instant they see you, be aware of how you move and act around them. Your body language consists of where you are in relation to your horse, your overall stance, and your movements. Your posture and movements can indicate, among other states of mind, confidence and strength or insecurity or aggressiveness. Since horses are followers, they are willing to be guided by a leader who walks with sure steps, has smooth movements, and whose heartbeat, breathing, and pheromones indicate confidence.

Although it is rarely appropriate to be aggressive with horses, there are situations where you need to be assertive. Other times a passive, nonthreatening manner is more appropriate and productive. Your body stance and movements will tell a horse whether he should be afraid, attentive, or relaxed.

Natural Aids

You use your natural aids all the time when you are around your horse or riding him. Your natural aids are your body, hands, mind, and voice.

Your Body: Weight and Balance

The body, usually considered to be the legs, seat, and back, is essential for communication through weight and balance when riding. During ground work, your body position and activity can encourage a horse to move forward, stop, turn, or stand still. Coordination of the body parts in harmonious gestures makes ground training and riding appear easy. All good trainers will tell you, however, that they have had to spend hours of trial and error to learn the appropriate choreography.

Your Hands: Guidance and Direction

Your hands and arms give a horse direction, both when you are riding and during ground training. In addition, we often

Coordination of your aids becomes especially important when you are also performing a task such as opening a gate.

use artificial aids to extend or intensify our hands and arms. Longeing whips, longe lines, riding crops, lead ropes, rope halters, and chains are all artificial aids that intensify the action of our hands. Artificial aids should never supersede a good understanding and use of the natural aids.

Your Mind: Evaluation and Decision

Your mind is your strongest asset when working with horses. You have the power to set the tone, choose the course of action, evaluate how things are going, reinforce behaviors, and modify or completely change your tactics. Your powers of observation and reason allow you to find a way to meet your goals while respecting your horse.

Your Voice: Instruction and Reassurance

Although horses don't verbalize the way we do, they do respond to our voices. It is appropriate and productive to use voice commands when working with horses, especially for ground training, and especially when you are the main person handling a horse.

The reason clinicians do not advocate voice commands more often is because most of them are talking constantly to the audience, and it would be difficult for a horse to distinguish a voice command out of all of that. Voice commands are not customary in most show ring settings but they can be a means to an end such as "Whoa" when a reiner asks his horse for a sliding stop. In most at-home training situations, voice commands are not only appropriate but also very effective.

Voice Commands

Horses can discern a wide variety of voice commands. They can quickly make an association between physical aids and voice commands. Eventually you can use the verbal cue alone to achieve the same response if you desire. This is handy for longeing, for example. If you follow a pattern with your voice

How to Speak Horse

"**Walk on!**" with a higher pitch on "Walk"; to start a horse from a standstill.

"**Ta-ROT!**" with a higher pitch on "Ta"; to trot a horse from a walk.

"**Waaaaaalk,**" in a drawling, soothing tone; to slow a horse to a walk.

"**Trrrrahhhhht,**" in a low pitch; to slow a horse to a trot.

"**Whoa,**" abrupt, low-pitched, with punctuated ending; to stop a horse promptly from any gait.

"**Eeeee-asy,**" in a soothing, drawn-out middle tone; to slow a horse within a gait or calm a horse.

"**Let's GO!**" or "**Can-TER!**" in an energetic, brisk tone; to get a horse to canter or lope.

"**Trot on,**" like "Ta-ROT!" but even; to get a lazily trotting horse to move forward energetically.

"**Baaaack,**" in a low, soothing tone; to back a horse during in-hand and long-lining work.

"**Tuuuurrrrrrn,**" in a melodic, descending pitch; to change a horse's direction when longeing.

"**Okay,**" a conversational prelude; to alert a horse that another command is coming.

"**Uh!**" a staccato command; to warn the horse to pay attention.

"**Goooood boy/goooood girl,**" spoken with pleasure and pride; to tell your horse he or she has done something well.

Adapted from *Longeing and Long Lining the English and Western Horse*, Wiley Publishing Inc., 1998

commands, it will prevent confusion. Voice commands should be consistent in the word used, the pitch, the inflection, and the volume.

Use Words Consistently

Commonly used voice commands include "Walk on," "Ta-rot," "Turn," "Canter" or "Let's go," then "Eeeeasy" and "Whoa." Which word you use for which command is more meaningful for you than your horse. Traditionally, "Whoa" or "Ho" is used for stop, but you could use "Bup" or "Stop" instead. To a horse, any word is fine as long as you use it consistently (no fair changing words or talking in sentences) and the word is onomatopoeic — that is, the word sounds like the action.

Horses are capable of learning quite a few words, but it is best to not choose words that sound alike. If you use both "Whoa" and "No" in his vocabulary, the horse is likely to stop whenever you say "No," even though you may have intended the command to stop another behavior.

Pitch Your Voice Appropriately

The pitch, or musical tone, of your voice can give a big clue as to the meaning of a command. Chirping a trot command in a crisp, high tone sounds like trotting, while a deep bass voice is more convincing for a "Whoa." When a mother tells her baby in loving singsong tones that he is "such a stinker," it is the tone of her voice that makes him laugh and coo, not the words she says. So it is with horses. The tone of your voice makes a statement of your mood and intentions more than the words you use.

Be Aware of Inflection

Connected closely with pitch is inflection or modulation: the rise and fall of your voice. A rising inflection tends to make a horse move forward, and a falling inflection to slow or settle. Thus it makes more sense to say "Can-*ter*," with a rising inflection on -*ter,* than it would to have a falling inflection on the second half of the word.

One of the biggest challenges for novice female trainers is finding an appropriate "Whoa" that is substantial and with a falling inflection. More often than not, the first attempts are more like "Whoa?" — which doesn't get your message across. Using a tape recorder is a great way to hear what you sound like as you are developing your voice commands.

Use Suitable Volume

Since a horse's sense of hearing is so keen, it is not only unnecessary but counterproductive to yell or even talk loudly to a horse. Horses can hear whispers. So in everyday handling and training, you and your horse might have quiet conversations that someone down at the other end of the barn or arena won't even hear. Loudness is rarely required. But for example, if a horse is zoned out and is rhythmically rubbing his tail on a fence, you might say "Twinkle, stop that" in a normal tone with no effect. That's when you might pump up the volume or add a loud clap or whistle to snap the horse out of his trance.

Learning

ALTHOUGH HORSES ARE NOT RANKED highly as problem solvers, their keen power of association and their adaptability make them extremely trainable. And they are *very* intelligent when it comes to being horses.

Horses are constantly learning. They adapt daily to their environment and respond as much to what we *don't* do as to what we do. That's why understanding how horses learn can make us more mindful and become better leaders.

The Brain

Man's brain comprises approximately 2.0 percent of his body weight, while a horse's brain is a mere 0.1 percent of his mass. When using brain size as an indicator of intelligence, it is easy to see why horses could be thought of as dumb beasts, but that is far from the case.

Parts of the Brain

The brains of mammals have similar parts. We cannot be certain as to the function of the various parts in horses, but can only speculate using research on the brains of humans and other animals as a guide.

Cerebrum

The thinking portion of the brain has four main lobes: frontal, parietal, temporal, and occipital. In the cortex, sight and hearing are processed and learning takes place. Deep in the cerebrum is the limbic system, where feelings are processed. Also in the limbic region is the olfactory lobe, which processes smell and taste.

Comparing Brains

Brain weight in vertebrates does not increase in direct proportion with body weight, as shown in the first list below. Other physiological factors help determine brain function and relative intelligence, ranked in the second list. Both lists are in round numbers.

Brain Weight as a Percent of Body Weight
Small bird 8.00%
Human 2.00%
Mouse 2.00%
Cat, dog 0.10%
Lion, elephant, horse 0.10%
Shark, hippopotamus 0.035%

Relative Intelligence Ranking
Human 7
Dolphin 5
Chimpanzee 2.5
Elephant 2
Whale 2
Dog 1
Cat 1
Horse 1
Sheep 1
Mouse <1

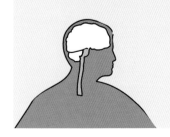

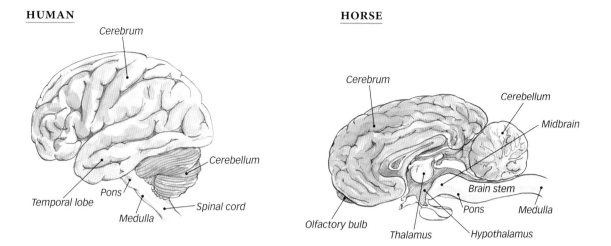

HUMAN

Cerebrum

Cerebellum

Pons

Temporal lobe

Medulla

Spinal cord

HORSE

Cerebrum

Cerebellum

Midbrain

Brain stem

Pons

Medulla

Olfactory bulb

Thalamus

Hypothalamus

Cerebellum

The athletic portion of the brain is located above the brain stem. The cerebellum regulates balance, coordination, and muscle activity. Conscious movement is a series of sequential events governed by neural activity in the cerebellum. Learned motor skills (training) occur here. Balance is governed in the cerebellum in conjunction with signals from the inner ear.

Brain Stem

The brain stem has four portions: the medulla, the pons, the midbrain, and the thalamic region. The medulla connects the spinal cord to the brain and is responsible for basic functions such as breathing, digestion, and heartbeat. The pons is where the balance between arousal and sleep is regulated. The midbrain is where memory is stored. The thalamic region is comprised of the pituitary gland, which controls hormone production related to sex and stress, and the thalamus and hypothalamus, which regulate temperature, hunger, and thirst and govern the endocrine and autonomic nervous systems.

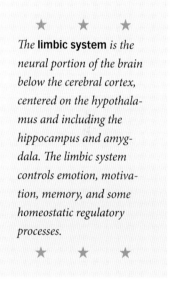

★ ★ ★

*The **limbic system** is the neural portion of the brain below the cerebral cortex, centered on the hypothalamus and including the hippocampus and amygdala. The limbic system controls emotion, motivation, memory, and some homeostatic regulatory processes.*

★ ★ ★

Mental Processes

As horses learn, they don't reason, they react. They are objective realists. Understanding how they learn is not only interesting, but also essential for communication and training.

Power of Association

Horses have a natural ability to link a stimulus with a response, the basis of classical conditioning. Although this is great for an experienced trainer, it can cause problems for a novice who makes mistakes — the horse is always learning, whether we want him to or not.

For example, suppose you are teaching your horse to back up in hand, and you face him and start walking toward him saying "Baack." At first he doesn't know what you want him to do. If you bump the noseband of his halter by tapping on the lead rope, he might take one step back. If you reward him by stopping, praising, and maybe rubbing him, you are training him using classical conditioning. The next time you repeat the lesson, he'll probably realize that when you walk toward him from the front and say "Baaack," you want him to move in reverse. This quick linking of action and reaction, stimulus and response, is one reason horses are so trainable.

If, however, a novice tries the same lesson and the horse lunges forward, rears, or veers sideways three or four times in a row, he has learned the wrong association. He has learned that when someone walks toward him, says "Baaack," and tugs on the halter, he rears and they quit bothering him. He has made an association that the novice didn't want, but it still was honest learning by association.

Anticipation

Once a horse has learned something, especially when it has been repeated too often, he might second-guess you and anticipate what you are going to ask. Although we may joke that such a horse is a mind reader or is going on "auto pilot," it's not really funny because pretty soon you have lost your means of communicating. Before you have even presented the cues, the horse is already beginning to perform what he expects you are going to ask — and he may guess right and he may not. Certain things can trigger anticipation, such as approaching

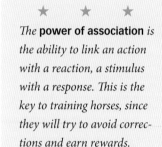

★ ★ ★

The **power of association** *is the ability to link an action with a reaction, a stimulus with a response. This is the key to training horses, since they will try to avoid corrections and earn rewards.*

★ ★ ★

a particular obstacle or place in the arena, or even your body language giving him a pre-cue.

You'll see this during longeing. Often a horse will sense when you are going to ask for a canter, for example, and he'll start just as you ask, not in response to your asking, so you might think, "Wow, he read my mind." In reality, he was reading your pre-cues. Although anticipation may seem harmless or even novel when first observed, it may develop into a habit that can make a horse virtually uncontrollable.

To prevent anticipation, you'll need to watch your pres and cues, so to speak. Vary the sequence of maneuvers, vary the location of lessons, and keep the lessons moving forward in a progressive fashion.

Memory

Horses are said to have a memory second only to an elephant's. If true, the horse is in distinguished company. Horses rarely forget lessons, good or bad. They remember past associations with alarming clarity and for long periods of time.

Once a horse learns a simple task, such as lowering his head, he'll likely remember it for months without use or review. If he learns a specific type of performance (reining, for example), he will remember it for years. After a long break, he might not be perfect the first time you ask him to review, but he will get back up to speed quickly and will be able to progress rapidly.

Similarly, we cannot erase a horse's bad memories: once scared or injured in a particular location, for example, he will be wary of that spot in the future. We can just add layer upon layer of new memories in hopes of burying the bad so deeply that they don't matter much any more and rarely surface.

Imprinting

The first type of learning that a foal usually experiences is imprinting. This process of species bonding between the dam and foal takes place during the first few hours after birth. The odors of the placental fluids and the sounds exchanged with the mare confirm innate behaviors in the foal.

Early handling by humans is a good idea. Once the mare and foal have bonded and all medical issues are resolved, the foal can be handled and touched all over. Since most foals are born at night, this training can begin as early as the next morning.

Imprinting occurs at birth when the mare and foal exchange sounds and smells. It is best if the human doesn't interfere during imprinting.

A band of horses follows its leader across a stream. A human can take on the role of that leader (see photos at right).

Modeling

Modeling or mimicking the behavior of other horses takes place in herds, so you may as well take advantage of the principle in training situations, too. When a band or herd crosses a creek, the herd members derive security from seeing the horses in front of them cross the water safely. Similarly, once a horse has bonded with his human handler, he can be more easily led across water, playing follow-the-human-leader. Young horses who observe other horses being tacked up, longed, and ridden seem to take to the process more easily than horses that live in isolation. That's why some trainers like to saddle up and work a group of young horses together in a round pen. Monkey see, monkey do, and there is safety in numbers.

However, modeling can work in negative ways, too. Some say tongue sucking, cribbing, wood-chewing, spooking, and being difficult to catch are socially contagious behaviors that horses obtain through mimicry.

Habituation

One of the first training principles you use when you work with a horse is habituation. This introduces a horse to a particular person, procedure, or object in order to gain his acceptance without fear. Related terms (in order from mild to extreme) are gentling, sacking out/desensitization, and flooding.

Gentling

Gentling is touching a horse on every part of his body and getting him used to all-over grooming. Although once a horse is not afraid of humans, he naturally loves to be rubbed on his forehead and neck, he must learn to accept and appreciate grooming elsewhere, especially in his ticklish and sensitive areas. (See How to Pat a Horse, page 38.)

Sacking Out

Sacking out is a form of systematic desensitization in which a mild stimulus is introduced at a low level, rest periods are given, and the stimulus is gradually increased. By repeated careful exposure to a certain stimulus, a horse's response can be diminished. Sacking out a horse with blankets and slickers is a way of gradually decreasing his apprehensions concerning the sight, sound, or feel of an object. If your end goal is to shake a noisy sheet of plastic over a horse's back and let it touch him, you would first rub him with a soft cotton blanket and gradually work up to the plastic over a few weeks.

Flooding

Flooding is exposure to full-intensity stimulus while restraining the animal until he stops reacting. With the above example, you would fully restrain the horse and then come at him from all sides with sheets of plastic, waving them wildly. Not only does this hold risk of injury to all parties, but it is an unnecessary means to an end.

A Safe Approach for Best Results

For safety, I prefer my horses to be sacked out but not totally desensitized, brain-dead, or robotic. When I am riding in the mountains, I want them to bring their instincts along. If I had removed all reflexes with aggressive flooding, it would be like riding a stuffed horse. I take care of my horses, and when we are riding I expect my horses to take care of me, but they could not react to danger if they had been numbed by flooding.

A beneficial use of desensitization is evident when your veterinarian gives your horse an injection. Often the vet will tap the injection site a few times with the back of his hand to stimulate the initial nerve firing before he inserts the needle. Thus prepared, a horse often doesn't react to the needle because his skin has been desensitized. A similar deadening occurs when you pick up a fold of skin and hold it for a few seconds before you insert a needle. The area around the site of injection has become dull to pain and the horse barely feels the needle.

I lead young Sassy into a mountain creek giving her the freedom to lower her head and look at the water. Then I mount up and ride her as she splashes across with great confidence.

Latent Learning

Learning that has been assimilated but has yet to be demonstrated is known as latent learning. Sometimes a horse has been taught a lesson but has not shown that he has really absorbed it. After a day or so off, however, the first time the horse is asked, he responds perfectly. This latent learning is commonly observed in horses. When it seems that a horse just doesn't get it, giving him some time to let it soak in often does the trick.

Learning Principles

All horses learn at different rates. You should have a training plan, but you will need to tailor it to each individual horse. In addition, horses are always learning. When you are feeding your horse, turning him out, or just grooming him, you are teaching and he is learning.

In order to learn what you want him to do, a horse must understand what you are asking.

At first, the request should be in simple terms, such as "I want to be able to touch you on the ribs without you being afraid." Then, over a period of many lessons, you can reach the stage where the horse learns to respond or not respond to various kinds of pressure. For example, he can learn the difference between leg pressure to encourage him to go forward, backward, or sideways, and pressure that is meant to drive him forward onto the bit and collect him.

Do Horses Know Right from Wrong?

Horses know instinctively that their behavior is "right" because it is what they were born with, their innate behavior patterns. To horses, everything they do is right — until we teach them otherwise. It is surprising how willing they can be to learn our right and wrong to get along with us. What a special gift. We can reciprocate by striving to be good and fair teachers.

Behavior Modification

A horse is always exhibiting some sort of behavior, whether he is peacefully grazing, pulling back while tied, or entering a trailer. Through behavior modification, you start with a base behavior and carefully mold the horse's actions into a safer and more useful pattern of behavior. You do this by linking a stimulus with a response following proven animal-training principles.

When a horse does what you want, and you wish him to repeat it in the future, you encourage that behavior. When he does something you don't want him to do, you discourage the behavior and show him a different way to act. Then you positively reinforce the new, desirable behavior. The more you learn about horses, the less correcting you will have to do.

Left Brain/ Right Brain

The left side of the brain governs scientific, logical, problem-solving issues, so it is often referred to as the thinking part of the brain. The right side of the brain involves pictures, patterns, emotions, and creativity, so it is referred to as the artistic side of the brain.

Horses tend to be more right-brain and people more left-brain. Humans can help horses develop the left brain, while horses can help people develop the right brain.

In order for your horse to understand your intent, whenever you work with him, your aids and reactions need to be immediate, consistent, appropriate, and concise.

Be Immediate

The timing of rewards (or punishment) is important. You have only a few seconds during or after the behavior to link it to the behavior. If you reward or punish before or after that time, you are reinforcing the wrong behavior!

For example, say you are turning out a yearling colt and just as you slip off his halter, he reaches over and bites you. Darn: he was rewarded with freedom right after he bit you. Bad as this is, it would make no sense to chase, screaming, after that horse; nor would it do any good to catch and punish him. If you did, you would either be chasing him like a predator or punishing him for being caught. Now, where's the sense in that?

You always need to "be here now" and pay attention when handling horses. You'll need to be aware of the nipping tendency in this horse so you can give him something else to do when you suspect he is thinking about biting. You need to be savvy and develop your horse sense to detect when he is thinking about trying this common, silly, adolescent behavior. When you read the signs, ask him to lower his head and back up, or return him to the barn, tie him to the hitch rail, and then turn him out a few minutes later.

To the horse, this is correct behavior. The approaching human would likely consider it misbehavior. Regardless of who is right, this horse's rearing can be eliminated using behavior modification.

Be Consistent

At first, if you stick with just one way of asking the horse to do something, it will be easier for him to learn. Once a horse masters the simple basics, you can start adding variations, which are essential for advanced riding.

For example, say you are teaching a horse to pick up his foot so you can clean his hoof, but his foot seems rooted to the ground. You use the technique of pinching the tendon, but it isn't working well for you with this horse. You try squeezing the chestnut and that works a little bit better, but it is still not getting a straightforward response. Next you try tapping the front of the hoof with the hoof pick and then nudging the heel with your boot toe; then you go back to the chestnut, or was it the tendon method? Well, I'm confused — I wonder how the horse feels.

Often, just as the light bulb is about to go on in the horse's mind, the human caves in and either quits or tries something else. If you are following a proven program, be consistent and persistent with your aids. The first time takes the most time; the next time will be much quicker. That said, I feel obligated to add (and I'll mention it again): if something is not working, things are escalating dangerously, and you can't resolve the situation, stop and change what you are doing or get help.

Be Appropriate

Choose an aid or cue that is appropriate for what you are trying to accomplish and use it with the appropriate intensity.

For example, when teaching a young horse to back up when you are opening a gate in-hand, it is appropriate to use the gate as a visual aid. You can position the horse to face the gate so that when you open it toward him, it will be a natural visual cue for him to start backing away from it. It would not be appropriate to slam the gate into the horse or bump him on the nose with it to get him to move.

Be Concise

Horses don't understand sentences, paragraphs, or long, drawn-out actions. The simpler your communication, the better. If you are teaching a horse to trot while longeing, a simple, crisp "Ta-rot" accompanied by appropriate body language, is ideal. "Come on there, Doofus. Move up there. Get going. Come on, Doofus!" is ineffective and counterproductive.

Behavior Modification Techniques

When you train a horse, you modify his behavior. Since horses have a strong power of association, they quickly learn what to do and what not to do with the proper use of stimulus-and-response conditioning. There are four ways to modify behavior: positive reinforcement, negative reinforcement, punishment, and extinction. All involve the use of stimuli or reinforcers.

Reinforcers, a.k.a. Stimuli

Animal behaviorists use the term reinforcer for the stimulus we apply to a horse to elicit a specific response and thereby train the horse. There is an action, and there is a reaction. If we choose our actions carefully, we will get the reactions we want from our horses.

Because using the words reinforcer and reinforcement (which comes later) can cause confusion, I'm going to use the word stimulus, but know that it and reinforcer are the same thing in the context of this book.

A stimulus can be a physical contact cue, such as pressure. It can be body language, such as stepping toward a horse. It can be a voice command that the horse has learned means something specific: for instance, "Eeeeasy". A stimulus can be something artificial, such as the rustle of plastic or the sudden appearance of a whip.

The horse interprets stimuli as being either positive or negative. Positive stimuli make a horse feel good; negative stimuli make a horse feel bad. Additionally, positive and negative stimuli can be of either primary or secondary type.

Primary positive stimuli are things that a horse inherently likes; he doesn't have to learn to like them. A horse is born liking primary positive stimuli such as food, rest, rubbing on the forehead, release of pressure, or having clear personal space.

Secondary positive stimuli are something that a horse learns to love because they are associated with a primary stimulus and give him an overall sense of well-being. When you use the praise "Good boy" with a rub, rest, or a treat, a horse learns to link the sound of the praise with a good feeling, so later just your voice alone can make the horse feel good.

Primary negative stimuli are things that a horse inherently does not like; he doesn't have to learn to dislike them. A horse is born disliking primary negative stimuli such as pain, pressure, and things that cause fear.

Rubbing the horse's forehead is a primary positive stimulus. The horse automatically loves it.

When you rub his forehead and say "Good boy," you are linking a secondary stimulus with the rubbing.

Later, when you say "Good boy" from a distance, your horse will remember the warm, fuzzy feeling of your touch and it will bring a sense of contentment to him.

These three drawings demonstrate linking of primary and secondary negative stimuli.
(1) Jerking on the halter is a primary negative stimulus. The horse naturally dislikes it.
(2) When you jerk on the halter and say "Quit" you are linking a secondary stimulus with the jerking.
(3) Later, when you say "Quit" from a distance, your horse will remember the physical discomfort of the jerk and he will be reprimanded by the secondary stimulus.

Secondary negative stimuli are things that a horse learns to dislike or avoid because they are associated with a primary negative stimulus and give him an overall sense of discomfort, unrest, or fear. When you scold your dinking (constantly fooling around) horse with a harsh "Quit!" and pair it immediately after with a sharp jerk on the lead rope, he learns to link the sound of the scolding with the upcoming bad feeling on his nose. Later, just your voice alone can make the horse stop his fooling around, pay attention, and stand still.

Positive Reinforcement

When a horse does something that you like, if you immediately give him something good or make him feel good, it will encourage him to repeat that behavior in the future. That is rewarding him, or using positive reinforcement. Your horse will be eager to repeat the behavior in the future because he likes what follows it. When rewarding, you can use primary or secondary stimuli. Rest and a rub on the forehead are primary positive stimuli.

Reward is the cornerstone of horse training, but take care that you don't inadvertently reward the wrong behavior. When that foal turns his butt toward you, and you laugh and think "How cute" and give him a good scratching over his tail, you are inviting problem behavior. He has been rewarded for turning his butt to humans and will repeat the behavior to get the same response. What's more, if you bring him in off the pasture as a yearling and he does the same thing, he will be frightened and confused when, now threatened by his bigger size, you

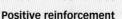

slap him sharply across the butt. So be aware of what you are rewarding, or it may come back to haunt you, supersize.

Another caution: food is the strongest primary positive stimulus, so be careful about what is happening at feeding time. If you go out to feed your horse in the pasture, he comes at you with ears back and threatening body language, and you just dump the feed and leave, you have just used reward to encourage him to behave like that again in the future.

Similarly, suppose one of your horses is the vocal type and his screaming is driving you crazy. If you give him extra feed to shut him up, not only have you contributed to his obesity, but you have also rewarded him for screaming and he will repeat the behavior in the future to get what he wants. (See more about this problem in Extinction, page 142.)

Negative Reinforcement

If as soon as a horse does what we want, we remove a negative or unpleasant stimulus, we have used negative reinforcement to strengthen the desired behavior. In the future, the horse will likely perform that behavior sooner so that the negative stimulus will be removed sooner.

For example, say you want your horse to move over. You apply pressure on his ribs with your hand or the butt end of a whip, or with your leg if you are riding. As soon as the horse begins to move to the side, away from the pressure, you remove the pressure by taking your hand, the whip butt, or your leg off the horse. His sideways movement has been encouraged using negative reinforcement and a primary stimulus. Pressure (an

★   ★   ★

Positive reinforcement *means giving a pleasant stimulus during or immediately after a behavior to encourage that behavior; also called* **reward.**

Negative reinforcement *means removing an unpleasant stimulus to encourage a behavior that is occurring.*

Reinforcement *means strengthening an association between a stimulus and a response. You can use primary stimuli (inherent) such as food or rest, or secondary stimuli (paired with primary and learned stimuli) such as praise or a pat.*

★   ★   ★

Don't punish a horse for an honest reflex. This horse might be reacting to the surcingle, cavesson, bridle, side reins, or whip.

unpleasant sensation) was removed when he moved sideways. He'll be more likely to repeat that behavior in the future, and it should take less pressure each time to get him to move over.

Another example takes place in the half-halt. A half-halt in dressage or a check in Western riding is a momentary gathering up of the horse. Using driving forces from the seat and legs and restraining forces via the hands on the bridle, the rider calls the horse to attention and asks the horse to gather up, compact, or collect for a second or two. As soon as the horse responds, the half-halt should be released.

In this case, as in many other aspects of training horses, the yield is much more important than the take. In other words, the result comes from rewarding the horse's compliance. If he responded the way you wanted but instead of yielding, you tried to hold him in that shape, it would give him no incentive to comply next time you asked.

Caution: Negative Reinforcement Can Backfire

When a horse bucks off a saddle or a rider, his bucking behavior has been strengthened by the principles of negative reinforcement. If the horse initially perceives a saddle or rider as something undesirable, threatening, or uncomfortable, and he succeeds in removing it from his back by bucking, then bucking behavior will likely be repeated in the future.

Punishment

Disciplining a horse immediately after an unwanted behavior can discourage him from repeating that behavior in the future. So when a horse acts badly and you do something that he perceives as unpleasant, you have punished him for his behavior. Some people are reluctant to discuss using punishment when training horses, feeling it is unfair and unnatural. All you have to do is watch a group of horses on pasture for a few days and you will witness some very real and harsh examples of punishment taking place.

When a newly weaned foal approaches an adult gelding and nuzzles his flank, the foal will be punished. It will be told in no uncertain terms to get away and stay away. It could be in the form of a kick or a bite, and the foal could be injured. That is natural horse behavior in action. When any horse approaches the feed of a more dominant animal, he will be punished. Punishment is a fact of life, a necessity in herds, and an integral part of "natural" horse training. If administered according to the principles in this book, punishment will be immediate, appropriate, consistent, and concise, and it will make for peaceful future relations.

For example, if you are leading your horse and he tries to run ahead of you or over you, and you give him a sharp tug on the halter, you are punishing him for his dangerous (bad) behavior. The pressure on his nose from the halter is a primary stimulus — he instantly knows he does not like the pressure or pain on his nose. You can link a secondary stimulus with this primary stimulus (jerking on the lead rope), if you also use a voice command such as "Quit." The horse will begin to associate your authoritative voice with controlling him, so that later your voice command alone will produce the same results.

Another example is an electric fence. When a horse leans over an electric fence into the next pen or pasture and he is shocked by the fence, he has been punished by a primary stimulus — he doesn't need to learn that a shock stings. He will be less likely to repeat that behavior in the future.

Punishment is something horses understand. Horses dish out severe and harsh punishment among themselves. The horse on the right is saying with tooth and hoof "Go away. Stay away."

Caution: wrongful punishment occurs when you punish a horse that does something right. Say you have a horse that is difficult to catch, and you are using the walk-down method to catch him. Finally, you get a halter on him, and once you do you give him a sharp jerk and say, "Next time don't be so hard to catch!" Well, if I were the horse, I'd be much harder to catch the next time, because I learned that when I finally stood still and let you approach and halter me, the first thing I got was punishment!

Extinction

When a horse has an undesirable behavior that we want to eradicate, sometimes we can use the principles of extinction to remove it.

Remember the horse that screamed when he wanted to be fed? He has learned that when he screams, he gets fed. You have inadvertently trained this horse to scream, using the principles of positive reinforcement or reward.

To rid the horse of that undesirable habit, you'll need to use extinction, followed by a new round of positive reinforcement. First, quit giving him something good (feed) when he exhibits the undesirable behavior (he is noisy). That's extinction. Then you must reward the behavior you want (a quiet horse). Once the horse is quiet, feed him.

If the habit is long-standing, it might take quite a while to change it, and at first, the horse might scream more loudly and add pawing and stall-door banging to the clamor, like a spoiled kid trying to wear down a parent in order to get candy. Just when it doesn't seem as though it could get any worse, the horse will likely try something different. When he tries to be quiet, that is when extinction is starting to work and you can start using positive reinforcement. Feed him when he is quiet and only when he is quiet. This particular example can be a real test of your commitment to training your horse.

Another example is a horse that has fought restraint and broken free; he has learned to repeat his fighting behavior through negative reinforcement. He removes something unpleasant (restriction) by fighting and overpowering a person or breaking equipment or facilities. Once he is free, he has learned what he needs to do in the future to get free — pull like there is no tomorrow! His behavior has been reinforced.

Don't cave in and feed your screaming horse or you will be training him to scream louder and longer in the future.

The Walk-Down Method

To use this means of catching a horse, start in a small pen and increase to a larger pen. Always walk toward the horse's shoulder, never his rump or his head. Never move faster than a walk. When the horse stops, scratch his withers. Always be the first to retreat. Eventually, halter the horse.

Case Study: Pulling Cure and Prevention

Here is how to use extinction with a horse that pulls when tied but has broken free only once or twice. Before using this method, set your horse up for success by working on several aspects of the problem.

Preparation

First, make sure he knows how to yield to pressure on his poll, including putting his head down and coming forward in response to very light pressure on the halter. Test him by doing rapidly paced in-hand work (such as walk, stop, walk, trot, stop, back) and seeing how he responds to the pressure. Test him further by ponying him (leading him from a seasoned, well-mannered riding horse) at all gaits. If he follows your pony horse like a butterfly on a string, proceed to the next phase, the extinction of pulling back when tied.

Procedure

In order to discourage fighting when tied, you must not allow the horse to gain his freedom. A common method is to tie a horse hard and fast to a safe, strong, high hitch rail or post with double rope halters and double lead ropes and let him pull until he gives up. Although this method can work, it can be dangerous. Here's another way.

Thread a 25-foot rope through a deflated inner tube (shown) or special tie ring set above the horse's head. Attach one end of the rope to the halter and hold the other end. When the horse pulls on the rope he will obtain a certain amount of slack, which you can then take up to return him to the starting point. The horse doesn't gain his freedom by pulling; he is just able to move somewhat, which keeps him from feeling trapped.

The horse's undesirable behavior is likely to get worse (pulling, rearing, scrambling, or falling down) before it gets better. Don't quit: this intensification is often a sign he is about to give in and the behavior is about to disappear and be replaced with another behavior, such as standing still. When he does stand still, he should be positively rewarded with rubbing, rest, and perhaps a walk around or release.

Caveat: Working with a confirmed puller is dangerous, and a cure might be impossible or just temporary. It is often best to leave it to an experienced trainer. Prevention is easier.

Sherlock's 25' lead rope is run through an inner tube. I'm holding the other end. When he pulls, I give him a little slack, then take it back.

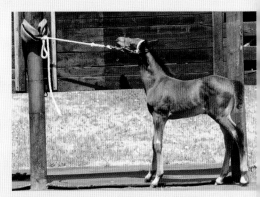

Now Sherlock is tied hard and fast to the inner tube. He tests it, feels pressure on his poll, and steps forward, yielding to the pressure.

Sherlock has learned that there is no escape and that standing tied is not only possible, but easy.

Fighting restraint can take many forms, such as pulling a leg away from a person who is trying to clean the hoof; breaking a halter, lead rope, or tie rail; and breaking equipment. If the habit is not too deeply ingrained, it is possible to use extinction to erase the fighting. However, if a horse has broken twenty-three lead ropes or halters when tied, he might uproot the entire barn or kill himself thrashing and fighting, before extinction works.

Repetition

Repetition is appropriate, necessary, and a friend to horses and to horse trainers, to a certain degree. But it is often carried to extremes. Certainly, horses respond favorably to consistent handling, and consistency involves repetition. Once a horse has been introduced to an idea and he responds even a little bit in the right direction, then it is a matter of building on that, and reviewing frequently afterward.

Repeating ground tie with Sassy in a variety of situations made her a ground-tie queen — solid, reliable, and confident.

Variety is the spice of life, and variety in horse training is good, too. It helps a horse develop into a well rounded, more confident partner and keeps him from anticipating and becoming robotic. Most horses would order up consistency with a little bit of variety.

In my experience, horses learn best if they are introduced to an idea, rewarded when they make attempts toward the desired behavior, and then given some time to let it soak in.

If a horse does not get it within a few attempts or a few minutes, either the approach needs to be changed or the horse or trainer needs a break. It simply makes no sense to continue repeating something with a horse when you are continually getting the wrong response.

Here is what is appropriate in regards to repetition: repeating something three to four times in a training session and holding training sessions five times a week. If you are working on backing your horse up (in hand or riding), the first time you introduce it, you might work on it for a few minutes until the horse starts to get the idea. Then you would leave it, work on something else, and come back to it a few more times during that first session. For every session after that, you would work on backing for a minute or so, three to four times in the session. Heck, if you do that, by the end of the week, you'll have a horse that backs promptly, straight, and in good position.

Here is what is not appropriate: repeating something a hundred times in a row. If you did this with the previous example of backing, you'd more than likely sour the horse to backing and he'd get sullen, resistant, and even lock up when you tried to back him. Why should he comply? All he has to look forward to is more and more backing. Knowing when to quit is an art. Read your horse. Be fair.

Shaping

Once a horse understands what you mean by an aid, you can then begin to ask for a gradual improvement in form, also known as shaping.

For example, when you are first teaching a tiny foal to lead, and you want him to stop, you might step in front of him, use a tug on the noseband of the halter, say "Whoa," or employ a breast rope to show the foal what you want. Your eventual goal is for the foal to stop square alongside you on a slack rope, just by cuing off your body language. But getting to that stage takes a series of steps that span several lessons. Each time the foal gets closer to the final goal, you should reward him by releasing pressure and gradually using fewer training aids.

In another example, when you are first teaching a two-year-old to canter on a longe line, you'll accept just about anything that leads to cantering without bucking or pulling. Your eventual goal is for the horse to canter directly from a walk, in proper balance, rhythm, and engagement and on the correct lead. Reaching such an advanced goal takes time, though. You'll likely see a lot of trot steps, wrong leads, breaking gaits, rushing, lifting of the head, and hollowing the back, but each time the horse shows improvement, he needs to be encouraged.

Shaping works best if you remember these principles: choose the best starting point, reward all good efforts, and don't move too fast toward the goal, but don't get stuck in a rut, either.

★ ★ ★

Shaping *is a term for the progressive development of the form of a movement; the reinforcement of successive improvements toward a desired behavior.*

★ ★ ★

ZIPPER'S SHAPING PROGRAM

★ Lope.
★ Continue loping.
★ Lope in a straight line.
★ Lope on the correct lead.
★ Lope in balance.
★ Lope with collection.
★ Lope from a trot.
★ Lope from a walk.
★ Lope from a halt.
★ Lope anywhere.

SHERLOCK'S LEADING PROGRESSION

When teaching a foal to lead, set yourself up for success by first leading the foal behind the dam. Then with the dam nearby, lead the foal using a butt rope to cue him to move forward. Later, lead the foal without the butt rope and out of sight of the mare.

Choose the Best Starting Point

In the foal-leading example, introduce "Whoa" when the foal is most likely to want to stop anyway. Have someone lead the mare next to the foal, and when the mare handler stops the mare, ask the foal to stop. It would be much more difficult to take the foal away from his mother the first time you asked for him to stop. You'd likely have a frantically rearing youngster.

When teaching the two-year-old to canter, you'll have better success asking for a canter when the horse is fresh and eager to move on than you will if you wait until he is tired.

Reward All Approximations to the Desired Behavior

With the foal, when you release the pressure on his lead rope when he stops, it is reinforcement. Letting the foal stop by his mother's side is also a reward. Even when a foal starts to slow down, you can give him a rub and say "Good" to let him know he is getting the idea. Soon the foal can learn that a soothing touch or word is a reward.

When you ask for a canter and the two-year-old bolts into a gallop, remain calm and praise the horse for his attempt to make an upward transition. Even if a young horse takes the wrong lead, at least he cantered, and it is best to reward him for that before you begin working on the leads. If you try to work on too many things at once, the horse will likely become confused. If you are fair and conscientious in what you ask of your horse, soon the horse will relax and the correct response will come easily.

Don't Go Too Fast Toward a Goal

If you try to reach perfection in just a few sessions, your horse might miss some valuable pieces of the puzzle or connections between the lesson's components. The beauty of a systematic training system is that when you have problems, you always have a progression that can be reviewed. If you move from point A to point D and skip B and C, then you have fewer places for review and fixing.

You might get the job done by being very harsh with the foal, and determined to get him to stop and stand still right from the beginning, but it will be through fear and physical pain rather than learning. This won't make the next session any easier — if you can catch the foal at all!

If you keep after the two-year-old and in one session expect him to canter on the correct lead in balance and in rhythm, you might be able to accomplish it with a very talented horse. Most horses, however, would become very tired, and you might do more harm than good. Remember, often the slower you go, the faster you get there, and the longer lasting the results are.

Don't Get Stuck in a Rut

After a while you may find that you get the job done pretty well but no longer gear the lessons to forward progress. This can make it more difficult to advance the horse's training later. Horses are creatures of habit, so be aware of what you do repeatedly with your horses. Keeping the lessons progressing will yield maximum performance and satisfaction.

If you lead a foal next to his dam for four months, then it will be more difficult to convince the foal that he can operate independently than if you had started the independent work in the first month of training.

For the first three months of longeing, if the two-year-old is allowed one or two trot strides in between a transition from walk to canter, it will be more difficult to erase them than if you started working on eliminating them after a couple of weeks.

The use of side reins is both an art and a science. They can be a valuable training aid but are sometimes used as a shortcut, bypassing important intermediate steps.

Training

NOW THAT YOU HAVE A GOOD IDEA of what a horse is made up of physically and mentally and you are familiar with the principles of learning, some general training guidelines are in order. Along with a bit of philosophy, I'm going to break my own rule about avoiding anthropomorphism and start out with a list of training rules from the horse's perspective — straight from the horse's mouth, so to speak.

I will then talk about setting goals, the phases of training, and the various types of lessons you can work on with your horse. Afterward, I will give you some guidelines for a typical training session, taking into account the instincts and behavior of the horse. With that, you're good to go!

12 Training Rules From the Horse's Viewpoint

If your horse could talk and you asked him how he'd like to be treated, he might answer something like this.

1 **No breaking.** Bend me, but don't break me. Present me with simple lessons that I can master, and build on those. Don't force me to change; invite me to change. Be calm and patient, and you will be amazed at what I will do for you.

2 **Be clear.** If you can tell me what you want me to do and I can do it, I will do it. If I don't understand you, don't punish me; ask me another way. I want to cooperate.

3 **Treat me like a horse.** I'm a horse and proud of it. Although we can be good friends, I'm not a person and I'm not your puppy, either.

4 **Be flexible.** I know you want me to master a certain action today, but cut me some slack if you see that I am distracted or tired or confused. Sometimes I need the reassurance of reviewing something simple that I already know.

5 **Focus, please.** You always ask me for my attention when we work together, so I'd like you to pay attention to what we are doing, too. Turn off the cell phone; forget about the sales contract on your desk or the fender bender your kid had or your recent medical results. Be here with me now, okay?

6 **Set the scene for success.** Since you know I am afraid of loose dogs or lawn mowers next to the arena, help me get over those fears first before you ask me to do something while those things are going on. Eventually, I'd like to be able to do anything, anytime, anywhere for you, but I have a lot of insecurities to overcome. With your help, we can do it.

7 **Be consistent.** When you are first asking me to do something, such as to put my head down so you can examine my ears, if you ask me the same way a few times in a row, I'll get the idea and, hey, no problem. But if you work with me a few times and then let your friend Joe handle my ears, and he does it really differently, I may get startled and he may get angry. It is going to be harder for me to figure out what I should do. If you are consistent until you see that I've got it, then you can start varying and adding. If you take your time, you will be surprised at all the variations I can learn. If I have trouble catching on, you can always review the first way I learned, which is locked in. Just give me a starting point and be consistent — I like that.

8 **Bond with me my way.** I like to be rubbed on my forehead and my neck; that makes me relaxed and content. Don't tickle the end of my nose or my flank or my belly, and please don't slap me hard, thinking I like it. Just use firm, circular, rubbing motions and we'll be buddies forever.

9 Take your time. When you are in a rush and move around me in a hurry, you smell anxious and I can sense your accelerated heart rate. Sometimes I get nervous and hyper, too. When you skip a step and ask me to do something new, sometimes I get lost and then can't remember the simplest task. I like it best when you move smoothly around me, letting me know what you are doing and taking as much time as it takes for us to figure it out together.

10 Be optimistic. When you walk toward me, I can tell if you are expecting things to go well or badly. If you are projecting a smile, I feel positive about working with you. On those days when you are in a rush or anticipate problems, I pick up on that and tend to shift into defense mode, because given the choice, I'd rather flee than fight. If you're happy, I'm happy.

11 Be fair and realistic. I really appreciate that you understand me, because then you won't ask me to do something that I am not physically capable of doing. You'd never ask me to carry or pull too much weight. And you'd never ask me to cross an impassable bog or go down a dangerously steep cliff. As long as you treat me fairly and only ask me to do reasonable things, I will never refuse you.

12 Be objective. When you and I are working together, report what you see, not what you interpret. When the back cinch strap hits my hind legs and I raise my leg, realize it surprised me. Since I can't see down there, my reflex is to kick at something that might be attacking my legs. Of course, when I have a minute to think about it, I realize that nothing is going to harm me, so I no longer lift my leg, but at first I just react. If you think, "Boy, you son of a gun, you are not going to kick at me!" and get mad at me, then we have a problem. Once you get to know me, you'll understand why I do certain things and give me the benefit of the doubt. In this way, you can help me overcome my fears.

Training Philosophy

If you are like me, you want to produce a horse partner that you can count on and one that can count on you. You want to feel safe working together, be able to communicate well, and enjoy your work. I've written about training extensively in other books (see Recommended Reading, page 178).

A Win-Win Partnership

As we move to the training principles section of this book, here's an overview, one I wrote for one of my very first books.

"When you face your horse in a training pen, neither of you wants a fight. And neither of you wants to be frightened or injured. You basically want to get along. If your goal is a long-term partnership, then you must reach an understanding and an effective system of communication. You must watch each other carefully, listen acutely, and respond honestly. You need to make the rules and be in charge. But, in order for the partnership to be successful, the rules should be based on the natural instincts and talents of the horse.

"For a human to win, it is not necessary for a horse to lose. You should not have to take things away from a horse or break him into fragments in order to train him; rather you should add to the horse. The goal should be making, not breaking.

"Horses, by nature, are generally cooperative and interested in developing an interaction with humans. So don't make the mistake of viewing a horse as your adversary.

"I hope that you have a good horse to work with and that you take your time and enjoy the experience — after all, isn't that the reason we all got into horses in the first place?"

— From *Making Not Breaking* by Cherry Hill
(Breakthrough Publications, 1992)

Training Goals

It is good to have goals but be flexible so you can adapt them to each of your horses each day. If you design your training program around a horse's natural behaviors, inclinations, and physical abilities, you'll have a better chance for success.

Designing an Effective Training Program

A training program is an individualized calendar of events that you design for your horse to suit his age, level of training, and his temperament. If it is well designed, it will suit the horse and it will help you accomplish your various subjective and objective goals. Think of a training program for horses as spanning weeks, months, or years, rather than minutes, hours, or days.

Subjective goals are those that you can't measure scientifically, such as a willing attitude, cooperation, trust, and respect. They are the foundation for the objective goals.

Objective goals usually involve performance of specific maneuvers, such as standing still when you mount, cantering on the correct lead, or clearing a four-foot fence. It is usually easy for you to see whether your horse has or has not met an objective goal. Eventually, the matter of form or quality of performance of objective goals enters the picture and then the quality of performance becomes your lifelong goal as a horse trainer. But never lose sight of the subjective goals. Never sacrifice your horse's trust or attitude to go higher or faster.

★   ★   ★

Training *is replacing a horse's inborn fear of man's world with trust and respect while preserving the horse's curiosity and willingness to learn.*

★   ★   ★

Zinger: willing, cooperative, trusting, trustworthy, respectful, and respected.

Goal Setting

The more you can specify what you want to accomplish, in your mind or on paper, the easier it will be to reach your goals. All styles of riding have the same basic goals that are taught in approximately the same order. Your list might look something like this.

My Goals for Basic Riding

☐ Mount with horse standing still

☐ Walk

☐ Halt

☐ Turn both ways

☐ Dismount

☐ Trot (walk to trot; trot to walk)

☐ Ride corners and large circles

☐ Turn on the forehand both ways

☐ Ride with other horses in the ring

☐ Ride out of the ring

☐ Serpentine, half-turns

☐ Canter (trot to canter, canter to trot) on both leads

☐ Square halts

☐ Back

☐ Walk around, turn on the hindquarters

☐ Negotiate obstacles, such as gates, small fences, and so on

Phases of Training

How you design the training program for your horse will depend on your horse's conformation, age, prior training and conditioning, and your (competition) goals and schedule. All training programs, whether casual or formal, tend to go through three phases: familiarization, learning skills, and improving form.

Phase 1: Familiarization

During early lessons, you need to accustom your horse to training routines, tack, and sensations. These include being touched all over, being groomed, having a saddle on his back, feeling the restriction of a girth, accepting the presence of a bit in his mouth, adjusting to the weight and sight of you on his back, tolerating the sweat during a session without being able to rub or roll, and following a daily work routine. In addition, you'll need to help your horse develop mental concentration so he can pay attention for progressively longer periods of time.

In the early part of a training program, these are the only goals and they are very important: the basics upon which all other training is built. Your horse needs to be relaxed and comfortable about all of these routines before you move on.

Phase 2: Learning Skills

Once you feel your horse is relaxed about the basics of riding or driving, you'll need to start teaching him specific skills.

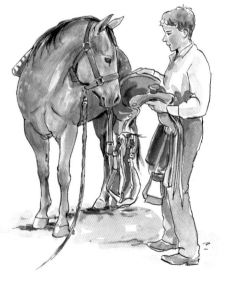

Phase 1 includes such basic routines as saddling.

Cultivate willing and correct responses to your aids whether you are on the ground or riding.

During the skills phase, there are many objective goals. For example, your horse needs to learn to move forward either in response to your body language or a longeing whip on the ground or from leg pressure when you are riding. The simplest form of the forward lesson is halt to walk, but going forward includes all upward transitions (walk-trot, trot-canter, and so on) and extensions within gaits (extended walk, extended trot, and so forth).

Your horse also needs to learn to stop and to slow down. You'll need to use proper body language from the ground and effective aids from the saddle to garner the walk to halt and all downward transitions (trot to walk, and so on) as well as collection within the gaits (for example, the collected canter).

Another basic lesson is to move away from pressure. You'll teach your horse to move his forehand or his hindquarters or his entire body away from your hand or leg. This lesson is needed for many exercises, including moving over while tied, turning on the forehand, and side-pass.

During this technical skills phase, your horse is learning what to do and what not to do. He learns a battery of responses according to your goals and his skill level. First, your horse will amass a repertoire of basic skills and then he will begin learning how to differentiate between similar aids for different responses.

Phase 2 involves learning specific skills, such as the sidepass.

Phase 3: Improving Form

After your horse has learned his basic technical skills, your next goal is to help him improve his form. In the form phase, you will gear the work so that your horse performs in a smoother, more balanced, collected manner. In other words, you will help him improve the quality of his work. First your horse learned what to do; now he will learn how to do it in better form. Phases 2 and 3 are not separate — you'll constantly shift back and forth between reviewing a skill and honing the performance. Your horse will usually tell you what to work on next.

No matter what the maneuver is, keep the important subjective goals in mind: forward energy, rhythm, suppleness, and relaxation, mental and physical acceptance of contact, relative straightness, balance, and precision.

Phase 3 focuses on improving form and the quality of the work, as in a collected lope.

As your horse adds new skills to his list of gaits and maneuvers, be sure you don't sacrifice these important basic goals just to learn a new exercise. Ask yourself:

☐ Is my horse still relaxed, loose, and supple when working on straight lines with forward energy?

☐ Is he keeping a regular rhythm at each gait? (It might be too fast or too slow at this point, but it must at least be regular.)

☐ Is my horse willingly accepting contact from my hands on the bridle and my legs on his sides?

☐ When I ask for more energy from the hindquarters, does he respond by moving forward?

☐ Does my horse track straight on a straight line?

☐ Does my horse feel balanced slightly to the rear, where he can work in a more collected form, rather than heavy on the forehand?

Physical Development

As you work your horse, regularly evaluate his physique. As your horse exercises and is trained, he will be using particular muscle groups more than others. With repetition, these muscle groups become dominant and subsequently affect his conformation. If the work you do with your horse is correct, he will develop an attractive, smooth, and functional physique. If the work is incorrect, his conformation could become imbalanced, with unsightly thick bulges in some places and hollow, weak spots in others.

Improving the Frame

Most horses start out with one of two basic frames. One is the horse with a shorter topline than underline: hollow back (sagging downward), high head, nose extended as much as 45 degrees in front of the vertical, croup higher than the withers, hind legs trailing. The other beginning frame is the horse with a long topline and a long underline: relatively flat, relaxed back; a low, flat neck; nose out at a 45-degree angle; heavy on the forehand. Both horses should be gradually developed so that they can strengthen and round their necks, backs, and croups; shift more weight back to the hindquarters; and increase the carrying capacity and activity of their hind legs.

If your horse is of the first type, he'd benefit from long, low work such as the posting trot, something that will stretch and

★   ★   ★

Self-carriage *refers to a form of posture and movement that is balanced, collected, and expressive, and that is either natural or developed and performed by the horse without aids or cues from the rider.*

★   ★   ★

Physical Goals

Keep these goals of physical development in mind as you work with your horse:

★ Gradually change a horse's flat or hollow topline to a bowed topline.

★ Develop suppleness and strength evenly on both sides of the body.

★ Gradually shift the weight of the horse from the forehand to the hind-quarters.

★ Improve the style or expression of the horse's movement.

★ Improve the quality of the gaits.

CHARACTERISTIC FRAMES

A high head and a hollow back

A flat topline and a low head

The goal, a balanced horse

relax his back. If you are starting with a horse of the second type, use only long and low work to warm him up and assure that he is supple. From there you can gradually introduce him to the idea of elevating his forehand and shifting his weight rearward. Upward and downward transitions between the halt, walk, and trot are a good means to initially get the horse to step under with his hinds and slightly elevate his front end. Training him to carry more weight on his hindquarters will take considerable time. It may take several months or more to see signs of improvement in overall carriage.

After several months of correct ground work or riding, however, your horse will likely begin to show signs of a slightly rounded back, a slight rounding of the neck, and a slight lowering of the croup. One of the most visible differences at this stage is that the horse can carry his nose comfortably and steadily at about 25 to 30 degrees in front of the vertical. His hind legs will step farther underneath his belly. It is beneficial to ride your horse in this type of frame for the next year of his training. He will likely show glimpses of self-carriage during that year.

Further developing your horse's frame is a process that is also accomplished by small degrees. At first, ride your horse in a slightly more engaged frame for only a short period of time (a few strides, a few minutes). Then allow him to return to his established level of self-carriage or give him a break on a long or a loose rein.

Ride your horse in an engaged frame for a few strides, then let him relax out of it. Pick him up again for a little longer. Gradually increase the time you ride your horse in a collected frame until it becomes second nature to him and he starts to develop self-carriage.

Content of a Training Session

Each time you handle your horse, think of yourself as a choreographer of events leading to your horse's finest and most advanced work. If you lay the foundation through carefully planned and executed exercises, you will have a better chance of helping your horse reach his full potential.

Types of Work

Choose the type of work that is most suitable for your horse's age and level of training: forward work, gymnastics, lateral work, or collected work. Horses generally tell you what you should work on. A horse will tell you when he needs a review or when it is time to move on to the next step. If you study your horse and tailor your training program to fit him, you will make great progress. If you try to cram a square peg into a round hole, well, it will be tough. Additionally, don't forget to work from both sides.

The Basics

Horses learn well using a progressive approach. If you start with something simple, something a horse can understand, master mentally, and perform well physically, you build up his confidence and interest. Then you can add the next step, and before you know it, you and your horse will be capable of performing all sorts of skills together. This is called starting with the basics.

What are the basics? They are the foundation on which all other lessons are based.

1. Don't be afraid
2. Show mutual respect
3. Pay attention
4. Move
5. Stop
6. Yield

While you are teaching the basics, take care of small problems before they become large ones that can't be fixed.

For example, if you start bridling your horse and find he raises his head, you will first have to teach him that he doesn't need to fear and avoid you by raising his head. If you find he is sensitive about you reaching for his ears and poll, you will have to desensitize that area first. When it comes time to put the bit in his mouth, you might find he does not like his mouth, lips, or teeth to be touched. You need to stop, go back to square one, and work on that. Often, if you spend a week or so taking care of the little things, you will find it improves your horse's behavior overall and will save you from constantly nagging over those little things in the years to come.

If there is a lesson that requires immediate attention, such as preparation of a new yearling for farrier care in two weeks, the lessons should be conducted several times a day so that steady progress is made. Shorter, frequent sessions yield much better and longer-lasting results than does one all-out marathon session.

Young Horse Sessions

If you are working with a suckling, weanling, or yearling, shorter, more frequent sessions are best. For a suckling or weanling, a session might entail grooming for five minutes, then fifteen minutes of in-hand work, followed by five minutes of grooming. The yearling might warm up with some in-hand work, learn the basics of longeing or ground driving in a twenty-minute lesson, and have five to ten minutes of grooming at the end of the session.

Forward Work

Forward work is energetic work with minimal constraint, something most horses naturally like. Examples are an energetic walk, trot, or canter in a straight line with very little contact from the bit and minimal bending or flexing. Forward work is always suitable for a warm-up and cool-down for any horse. In the very early stages of training, especially of a young horse, forward work might be the only suitable type of work for the entire training session.

Gymnastics

Gymnastics add contact, bending, and changes of rein to the forward work. Examples are circles of various sizes, half-turns (reverses), serpentines, and change of rein across a diagonal. Even though you've now added bending, the movements still proceed on one track, with your horse's hind feet following in the footprints of his front feet, whether on a straight line or a circle. In other words, there is no lateral work yet, which would offset his hindquarters to his forehand.

Gymnastics might constitute the new work portion of a young horse's lesson, and this activity is appropriate for the

Here is an example of forward Western ground driving.

Circles are the key to gymnastic work.

last part of a warm-up and for the review periods for a more experienced horse.

Lateral Work

Lateral work begins the intermediate phase of training and includes figures that have a sideways component to them. Examples are turn on the forehand, leg-yield, turn on the hindquarters, leg-yield, side pass, half-pass, and so on. In most cases, lateral exercises are worked on after a horse is warmed up and has done ample forward work.

Collected Work

Collected work aims at developing your horse's longitudinal (front-to-back) flexion and strength and balance in his hindquarters. Examples include all upward and downward transitions, such as walk-trot, trot-walk, trot-canter, canter-trot, walk-canter, canter-walk, and backing. Generally, this work is the focus of new work for an intermediate to advanced horse.

In a half-pass, a horse moves sideways and forward at the same time.

Collection develops and requires strength and balance.

A Typical Training Session

No matter what stage of training your horse is currently in, try to pay attention to how you organize each training session. A typical session, whether ground training or mounted training, includes preparation of your horse, warm-up of you and your horse, the training session, cool-down, and post-session care of your horse.

Preparation of Your Horse

Your relationship with your horse begins with the first step you take toward him in the stall or pen. As you handle your horse during leading and grooming, be direct and precise in your body language.

Tie your horse in a comfortable manner that confines his movement to an area that makes him safe for you to work on while you are grooming and tacking. Be very matter-of-fact as you perform routine tasks such as cleaning his hooves, applying fly spray, clipping, and so forth. Your horse's acceptance and cooperation during these common practices affect his mindset and attitude for the upcoming session. Use grooming to stimulate your horse or to relax him, depending on his nature and the stage of his training.

Warm-Up

If your horse is bridled as you lead him to the training area, remember that leading a bridled horse is different than leading a haltered horse (see box on opposite page).

Catch and halter your horse.

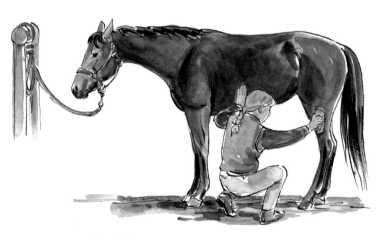

Groom and tack him.

When you reach the arena, stop your horse straight and square and give him your command to stand still. Then take your time as you prepare to start the session: this will develop patience in your horse.

For example, if you are going to longe, take plenty of time organizing your longe line or fiddling with your sunglasses or hat — anything to let your horse relax and stand still and not anticipate. If you will be riding, check that the saddle is straight and then step in front of your horse and see if your stirrups and bridle are even. Give the girth its final tightening. Put on your gloves and sunglasses, secure your hat or helmet, and mount. Sit for a moment without doing a thing.

When you decide it is time to move off, whether longeing or riding, give your horse the appropriate signals. Follow this routine consistently and your horse will not develop the habit of dinking around or walking off as you mount.

A warm-up serves many purposes for both you and your horse. It readies the neurological pathways, alerting them for signals, thereby increasing coordination during the more demanding work that will follow. It increases blood flow to the skeletal muscles, which increases their strength of contraction and allows them to stretch without damage.

Stretching exercises, such as an active long trot, should not be used as the first part of a warm-up as they may result in torn

Lead him to the arena.

Leading with Halter vs. Bridle

Leading a bridled horse requires a different method than leading one wearing a halter. With a halter, the lead rope communicates with the halter ring or knot under his jaw and you direct him with left, right, and backward movements of the rope. This exerts pressure on the cheekpieces, noseband, and crownpiece of the halter.

With a bridle, the reins are attached to a bit, putting pressure on the bars, tongue, and corners of the mouth. If you treat the reins of a bridle like a single lead rope, you would give confusing and contradictory signals to your horse's mouth. Separate the reins with your index finger and use the reins independently to indicate to your horse whether he should turn right or left or slow down.

Mount up.

fibers. It is always beneficial to move your horse at a walk for at least two to three minutes before starting a trot.

An energetic, forward trot at a slow rhythm is suitable for the end of the warm-up period; an explosive fast trot or complex work is not. Your horse will tell you when he is relaxed and ready to move on to the next phase when he shows you these signs:

⭐ Blows (exhales gently or forcefully through his nose)

⭐ Breathes long and deep

⭐ Mouths the bit, chewing and licking

⭐ Begins lowering his head

⭐ Reaches forward with his neck and head

Almost any horse improves after a warm-up. It takes the edge off a fresh horse and puts him in the mental state to pay attention and work. If you have a lazy horse, a warm-up will get his blood flowing and he'll be more physically stimulated to work. If your horse is hyperactive or hot, a warm-up will "smooth" him out: that is, his neuromuscular responses will start firing with more control. But be careful you don't use up all of your horse's energy in the warm-up. Save some for the actual training session.

A blowing horse, indicating readiness for work

Ideal Length of a Training Session

Remember, all handling and riding is training.

Age	Length of Session	Frequency
Foal	15 minutes	5 times per week
Weanling	30 minutes	5 times per week
Yearling	30–60 minutes	3–5 times per week
2-year-old	60 minutes	4–6 times per week
3-year-old	up to 90 minutes	4–6 times per week
4-year-old	up to 2 hours	2–6 times per week
5–20-year-old	1–6 hours	2–6 times per week
21 and older	30–90 minutes	4–6 times per week

The Heart of the Training Session

To custom-design a training session to fit your horse, try to think like a horse as you fit the pieces together. It is usually more appealing to and more productive for a horse if you break the session into several short sections rather than approaching it as one long block of time. For a one-hour session, you might use ten minutes for a warm-up and save ten minutes for a cool-down at the end. That leaves you forty minutes for the work. So the sixty minutes would go something like this:

The work session

★ Warm-up (long trotting): 10 minutes

★ Review work (something your horse knows very well): 10 minutes

★ Break (let your horse stretch and blow and relax): 2 minutes

★ New work (something your horse is in the process of learning and you want to work on together): 15 minutes

★ Break (a little longer break after the hard stuff): 3 minutes

★ Review work (go back to working on something that your horse knows very well and really enjoys): 10 minutes

★ Cool-down (some long trotting and walking on a loose rein): 10 minutes

Review

When you are ready to begin the review portion of the session, choose something your horse knows well and is capable of performing with relative mental and physical ease. Your horse will appreciate the mental break. He will be confident because he can do circles and serpentines at the walk, or a quiet trot in his sleep! But don't let him fall asleep while doing them — review work should be active, forward, lively, and on straight lines (no lateral work). Keep it simple.

Break

When you take the rest break, take care not to "throw your horse away" all of a sudden. Abrubtly releasing the reins or letting your horse collapse on the longe line just teaches him to travel heavily on his forehand. Let him, instead, gradually stretch down. Feed the reins or line out to him slowly until he is just moseying around the arena, blowing through his nose.

If a horse reaches down, stretching the top muscles of his neck, it indicates the preceding work was done correctly. If instead your horse throws his head up and bulges the underside of his neck out, it means his back is sore and the previous work was incorrect and tense.

As the rest period comes to a close, drive your horse forward with your lower legs or with your longe line and body language, slowly regaining the previous level of contact until you have your horse working in the state he was before the break.

Taking a break on a loose rein

New Work

At first you might want to use a watch to get a sense of how long you are taking for the various portions of a session, but before long you'll learn to read your horse's signs and know when it is time to move on to the next segment of the session. The new work period should occur during your horse's mental and physical peak, when he is tuned in and warmed up but not tired out. The new work is the more challenging work that your horse is learning, such as a series of precise transitions or advanced lateral work.

It is not so difficult to *start* new work. The hard part is knowing when to *end* new work. Knowing when to quit is both an art and science. Although you'd like to make a major breakthrough during every single training session with your horse,

How Much Is Too Much?

It is not what you use but how you use it that determines whether something is appropriate or excessive. Whether we are talking about time, tack, aids, or repetition, often less is more.

★ If something is not working after considerable repetition, stop and reevaluate.

★ When choosing whether you should ride with a snaffle bridle, bosal, bitless bridle, curb bit, or spade bit, remember that the improper use of any item of tack can result in mental abuse or physical injury. Your hands have the ability to turn a snaffle bit into a cruel instrument, or a spade bit into a delicate means of communication.

★ Longeing can be used as an enjoyable, productive interaction or as a way to exhaust a horse and deplete his spirit reserves.

★ Halters, ropes, chains, whips, and other ground-training tack can be used effectively and appropriately, or be intimidating and counterproductive.

The new work can include obstacle work, which most horses find interesting and engaging.

that isn't a realistic expectation. Sometimes it is necessary to quit while you are ahead. If your horse has given you an honest effort but begins to tire and make mistakes, he is telling you that it is time to move out of the new work period, even if you have not accomplished *your* goal of the day. When your horse starts making mistakes, he could be telling you that he has lost his concentration, he is past his peak, or he is just plain tired.

On the other hand, if your horse is still fresh and is being testy or stubborn, and in spite of your good efforts is not paying attention, then you may need to work him through a difficulty before you close the new work period. If you end the session in frustration, just think of how confused your horse must be. Try to find a satisfying point of closure for both you and your horse.

In many instances, it might serve you both better (and avoid a fight) if you walk on a long rein for a few moments, arrange your thoughts and strategy,

Peak Performance

Every horse peaks, mentally and physically, at a different time and for a different length of time. When a horse peaks, he is at the top of his physical game and he is alert and responsive. Things go smoothly.

At first, the peak might only last a few minutes. With any horse, the peak period will increase as his condition improves. You need to learn how to recognize and then predict your horse's peak so that you can make the most of that important zone. That's when you want to perform the new or most complicated work with your horse.

If you insist that a horse work too far past his peak, you risk destroying what you have gained in the previous work. You always want to quit the demanding work at a point where it is still fun and engaging for both of you, so that you can't wait to work on it during the next session.

and then return to the work. If your horse is in good physical condition and you know his level of mental concentration, you will likely be able to recognize those times when he is going so well that you can ask for more complex work. It is important to realize that when introducing new lessons, there always is the potential of running into a problem. If you make an effort to think things out ahead of time, you will be more capable of quickly deciding whether you should resolve a problem now or tactfully close the session instead and deal with the problem next time.

Break

The break after the new work can be a little bit longer than the previous break, but it should follow a similar format. Your horse will really appreciate this break.

If he shows his appreciation of the loose rein by stretching his neck down and reaching out with his nose, that means your work together has been well done. If instead he raises his head and hollows his back, that tells you that you have been holding him in tight position and when you let go, he had to counteract those muscles with a reverse stretch. This is not a good sign.

When you take contact with your horse through the bridle, seat, and legs, you will start the final work session, which is a valuable review.

Closing Review

Before the final review period, think of the problem areas that popped up in the new work so that during the closing review, you can touch on the basic principles that underlie them. Don't be tempted to work on new movements per se, or you may open a can of worms just when you are ready to quit. Instead, take your horse through a review of simple, basic principles so that during your next training session, he will have a better chance of performing the new work correctly.

For example, if your horse was stiff when you cantered left, don't canter left during the review period. Rather, run through some gymnastic exercises in both directions at the walk and trot to loosen his poll, throatlatch, neck, shoulder, rib cage, and hindquarters. I've often found that working to the right improves work to the left, and vice versa.

To preserve your horse's self-esteem and interest in his work, end the review session with something he does really well. It is good for you, too, to end with a good feeling about your training and riding, so as you wind down the session, think of several parts of the lesson that went particularly well. If you both end on a positive note, you will look forward to working together again soon.

Untacking

The Cool-Down

After a vigorous training session, it is important to let your horse gradually and systematically wind down from his work. Save at least ten minutes for this. The cool-down begins after the last review period in the session when you give your horse some slack in the reins, and ends when you have returned to the barn to untack.

A cool-down does not have to consist entirely of walking around on a long rein, although it can. Unless your horse is unfit, a cool-down can include some trotting freely on a long rein. That sort of loose, relaxed

Cooling out at the hitch rail

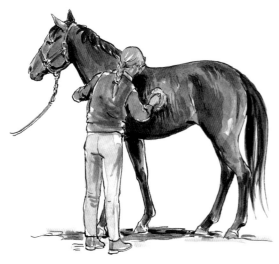

Grooming

Turnout

trotting will help flush accumulated lactic acid from the dense muscles of his hindquarters and will also clear his mind. After particularly hard work, you may choose to dismount, loosen the cinch, and lead your horse around the arena for the last five minutes.

If your horse is very hot, don't let his thick muscles cool out too quickly. Keep his back and loin covered with a quarter sheet or wool cooler so his muscles will dissipate heat more slowly.

Post-Session Care

What you do after the training session can have a lot to do with how your horse will greet you the next day. If you spray a hot horse with cold water, it can feel unpleasant and cause his muscles to become stiff. Besides, using water to hose sweat and dirt off your horse every day is not a good long-term management practice. A daily transition from wet to dry can be extremely damaging to the structure of your horse's hooves. Also, fungus and skin problems can occur when horses are frequently wet down and aren't allowed to dry thoroughly.

Most horses enjoy a vigorous rubdown with dry towels or burlap much more than a spraying with water. After wiping your horse down and applying a full cooler or scrim, you can tie him out of drafts in the winter and out of direct sun in the summer so he can dry comfortably. When your horse is thoroughly dry, curry and brush him vigorously, use a vacuum if desired to remove any dried sweat or shedding hair, cover him with a sheet, and return him to his stall or pen. As you turn him loose, give him a good rub on the forehead and neck and tell him he did a good job.

Epilogue

The dawn horse, also known as eohippus or hyracotherium, was the first equid to appear on earth 55 million years ago. Although DNA evidence is still coming in, for purposes of this discussion we can say that the modern horse, *Equus caballus,* has been domesticated for approximately 5,000 years.

To put the horse's relationship to man in perspective, come and stand in the middle of my 100 × 200 foot arena with me. If you start by looking at the in-gate and then let your eyes travel 50 feet to the first corner, then 200 feet down one long side, 100 feet across the far end, 200 feet up the other long side, and then 50 feet back to the gate, your gaze will have traveled 600 feet, or 7,200 inches. That represents the 55 million years of the horse's evolution.

Now, look at your little fingernail. It is approximately ½-inch wide. That represents the number of years that the modern horse has been domesticated.

I'm continually amazed at the adaptability of the horse, the number of things the horse is willing and able to do for man. But deeper yet, is my admiration for the spirit and nature of the horse.

Whenever we work with one of these special animals, the more we can think like a horse, the better for us all.

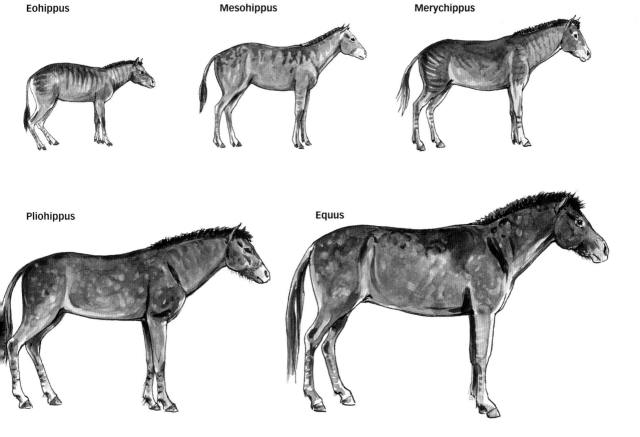

Eohippus

Mesohippus

Merychippus

Pliohippus

Equus

Parts of a Horse

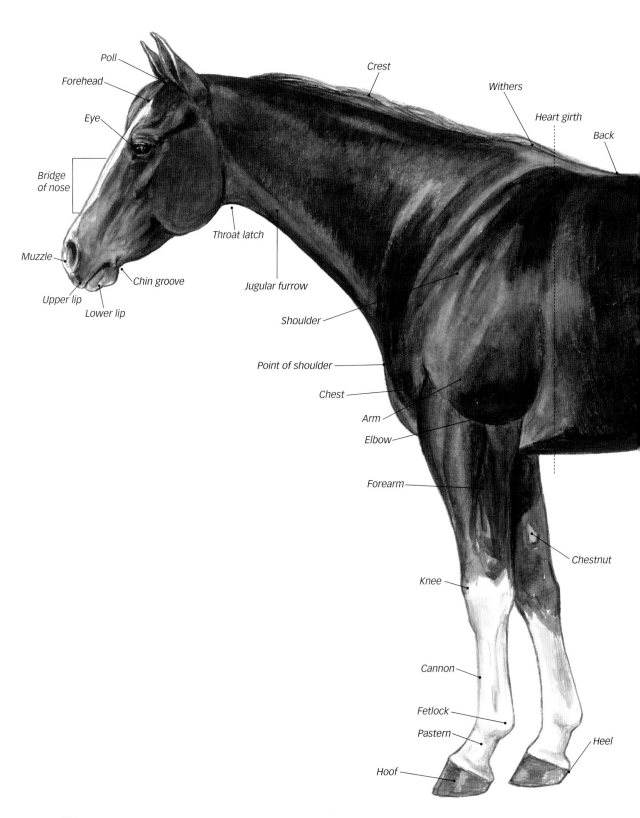

Poll

Forehead

Eye

Bridge
of nose

Muzzle

Upper lip

Lower lip

Chin groove

Throat latch

Jugular furrow

Shoulder

Point of shoulder

Chest

Arm

Elbow

Forearm

Knee

Cannon

Fetlock

Pastern

Hoof

Crest

Withers

Heart girth

Back

Chestnut

Heel

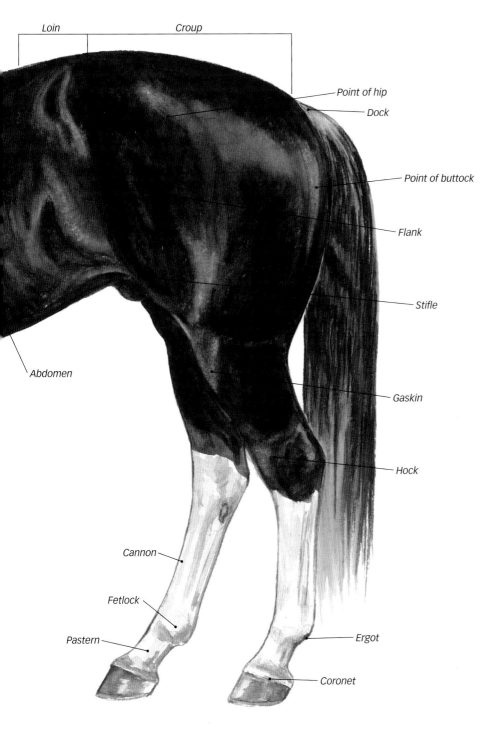

Loin

Croup

Point of hip

Dock

Point of buttock

Flank

Stifle

Abdomen

Gaskin

Hock

Cannon

Fetlock

Pastern

Ergot

Coronet

Body Language: Develop Clear Signals

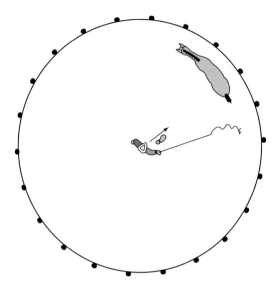

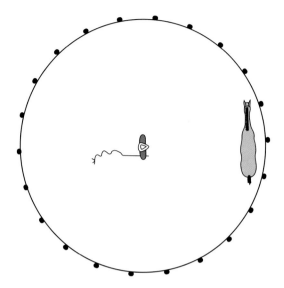

Drive: Simultaneously, step toward the horse's hindquarters with your driving foot (right foot when horse is tracking left); raise the whip; say "Walk On".

Neutral: Hold whip behind you with both arms at your side; weight evenly distributed on both feet; lower your gaze and take a deep belly breath.

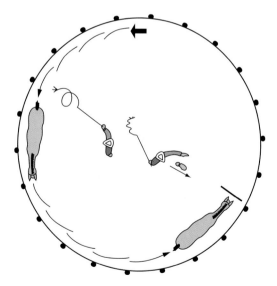

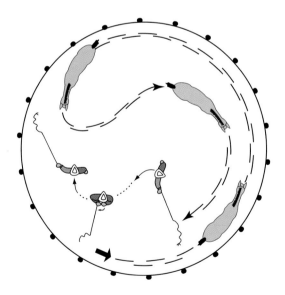

Stop: Simultaneously, step toward the horse's fore-hand with your blocking foot (left foot when horse is tracking left); lower the whip behind you; raise your left arm and say "Whoa."

Turn: With horse tracking left, start backing up and to the left while handing the whip behind you from right to left hand. Step to the left, raise the whip and say "Turn."

Glossary

aberrant behavior. Abnormal behavior. *See also* stereotypies

accommodation. The self-adjustment of the lens of the eye for focusing on objects at various distances.

acuity. Keenness or sharpness of vision.

adaptation. A change in behavior to conform to new circumstances; in vision, the power that the eye has of adjusting to variations in light.

agonistic behavior. Social interactions that serve to maintain order.

aid. The means by which a trainer or rider communicates with the horse. Natural aids are the mind, voice, hands, legs, body (weight, seat, back); artificial aids include the halter, whip, spurs, chain.

anthropomorphism. Attributing human characteristics to nonhumans.

anticipation. A response that begins before an expected stimulus.

attitude. A temporary behavior reflecting specific conditions.

bad habit. Undesirable behavior in response to handling or riding.

balk. Refusal to move.

band. A small, stable group of horses; in the wild, a breeding band consisting of mares is called a harem. A bachelor band consists of all male horses.

barn-sour. Describes a horse that has developed the bad habit of bolting back to the barn upon release. *See also* herd bound

bars (mouth). The bony, flesh-covered space between the incisors and molars where the bit lies; also called the interdental space.

behavior modification. A method of changing existing behavior.

binocular. Using both eyes at the same time.

buddy bound. Describes a strong bond between two horses that can result in separation anxiety.

check ligaments. Part of the stay apparatus of the forelimbs.

clacking. A submissive gesture of foals with lowered head and repeated opening and closing of mouth; also known as snapping or champing.

classical conditioning. Linking a stimulus with a response for the purposes of training.

claustrophobic. Uncomfortable and frightened when confined or crowded.

cold-blooded. Horses whose ancestors trace back to heavy war horses and draft breeds. Characteristics might include more substance of bone, thick skin, heavy hair coat, shaggy fetlocks, and lower red blood cell and hemoglobin values.

colic. Abdominal pain.

colt. A young uncastrated male horse usually between the ages of weaning and gelding (or stallion selection).

cue. A signal or composite of trainer aids that is designed to elicit certain behavior in a horse.

dam. The mother of a horse.

dominance hierarchy. Pecking order; social ranking of individuals within a group.

epimeletic behavior. Caregiving or attention.

epiphysis. The growth plates at the ends of long bones.

Equus. The genus of the horse; the species of the modern horse is *caballus.*

et-epimeletic. Soliciting care or attention.

evasion. Avoidance of an aid; for example, a horse that overflexes or gets "behind the bit" to keep from accepting contact with the bit.

extinction. Removal of a pleasant reinforcement to discourage the behavior that is occurring.

filly. A young female horse.

flehmen response. A behavior in reaction to a smell; the horse raises the head and curls back the upper lip, sending scent into the vomeronasal organ.

flight. Escape or running away.

flooding. An intense, overwhelming form of habituation. *See also* habituation

foal. A young male or female horse, usually under a year old.

gelding. A castrated male horse.

gregarious. Social, living in herds.

habituation. Repeated exposure to a stimulus, thus decreasing the horse's response to it.

herd. A large group of horses of mixed ages and sexes.

herd bound. Separation anxiety exhibited by an individual when he is removed from the herd, and may result in the bad habit of bolting back to the herd upon release.

hot-blooded. Horses whose ancestors trace back to Thoroughbreds or Arabians. Characteristics might include fineness of bone, thin skin, fine hair coat, absence of fetlocks, and higher red blood cell and hemoglobin values.

imprinting. The rapid learning in a young horse's critical period (first few hours of life) that reinforces species behavior and creates bonds.

infrasound. Any sound with a frequency below a human's audible range of hearing (i.e., less than 20 Hz).

instinct. Inborn, intrinsic knowledge and behavior.

intelligence. Ability to survive in or adapt to the human's world.

intermittent pressure. Application and release of an aid, in contrast to steady pressure.

latent. A type of learning that has been assimilated but has yet to be demonstrated.

limbic system. Neural portion of the brain below the cerebral cortex, centered on the hypothalamus and including the hippocampus and amygdala. It controls emotion, motivation, memory, and some homeostatic regulatory processes.

long yearling. A horse in the fall of its yearling year; usually 18 months of age.

meconium. A dark, sticky fecal material that accumulates in the fetal intestines and is discharged at or near the time of birth.

memory. Ability to remember previous experiences or training.

mimicry. Allelomimetic behavior, or copying the behavior of others.

modeling. Observational learning or mimicry.

monocular. Using one eye to see.

mutual grooming. Reciprocal nibbling along the neck, withers, and back between two horses, usually bonded buddies.

nasal turbinates. Passageways from the nostrils to the lungs.

near side. The left side of the horse.

negative reinforcement. Removing an aversive stimulus to encourage a behavior that is occurring.

nomadic. Wandering or roaming.

off side. The right side of the horse.

olfactory. Pertaining to the sense of smell.

pair bond. Two horses that exhibit a preference to stay together, sometimes so strong that it causes problems.

papillae. Folds and projections on dorsal surface of the tongue that contain taste buds.

pecking order. Caste system or social rank.

pheromones. Chemical substances secreted by an animal that elicit a specific behavioral or physiological response in another animal of the same species.

poll. The junction of the vertebrae with the skull; an area of great sensitivity and flexion.

positive reinforcement. Reward; giving something pleasant to encourage a behavior that is occurring.

power of association. The ability to link an action and a reaction, a stimulus and a response. The key to training horses, because they will try to avoid mistakes and earn rewards.

proprioceptive sense. The ability to sense the position, location, orientation, and movement of the body and its parts.

punishment. Administering something unpleasant to discourage a behavior that is occurring.

reciprocal apparatus. Part of the stay apparatus of the hindlimbs.

reflex. An unlearned or instinctive response to a stimulus.

reinforcement. Strengthening an association; with primary stimuli (inherent), such as feed or rest, or secondary stimuli (paired with primary, and learned), such as praise or a pat.

REM sleep. Rapid eye movement sleep; a stage in the normal sleep cycle during which dreams occur and the body undergoes various physiological changes, including rapid eye movement, loss of reflexes, and increased pulse rate and brain activity. Also called paradoxical sleep.

resistance. Reluctance or refusal to yield.

restraint. Prevention of acting or advancing by psychological, mechanical, or chemical means.

sacking out. Gentling, usually by accustoming to flapping objects.

seasonally polyestrous. Multiple breeding periods during a specific breeding season each year.

self-carriage. A form of posture and movement that is balanced, collected, and expressive and that is either natural or developed, and performed by the horse without aids and cues from the rider.

separation anxiety. Nervousness when bonded individuals cannot touch or see each other; can cause barn-sour, buddy-bound, or herd-bound behaviors.

shaping. The progressive development of the form of a movement; the reinforcement of successive approximations to a desired behavior.

slow-wave sleep. A state of deep, usually dreamless sleep that is characterized by delta waves and a low level of autonomic physiological activity; also called non-REM sleep, or orthodox sleep.

socialization. Development of an individual and his behavior through interaction among others of the same species.

spook. A response of jumping and running when encountering a frightening object or situation.

startle response. Spooking in place.

stay apparatus. A system of ligaments and tendons that stabilizes joints and allows a horse to stand with very little muscular effort.

stereotypies. Aberrant behaviors repeated with regularity and consistency. Examples are cribbing, pacing, and self-mutilation.

stress tolerance level. The point at which a horse can no longer absorb stress (noise, exercise, or trauma), and erratic behavior results.

substance. Of solid quality, as in dense bone or large body size.

suckling. The nursing foal.

sullen. Sulky, resentful, withdrawn.

supple. Flexible.

temperament. The general consistency with which a horse behaves.

ultrasound. Any sound with a frequency above the human's audible range of hearing more than 20 kHz).

vices. Undesirable behavior patterns that emerge as a result of domestication, confinement, or improper management.

voice command. A natural training aid that must be consistent in the word used, tone, volume, and inflection.

walk-down method. A means of catching a horse. Start in a small pen and increase to a larger pen. Always walk toward the horse's shoulder, never his rump or his head. Never move faster than a walk. When the horse stops, scratch his withers. Always be the first to leave. Eventually, halter the horse.

weaning. Separating the foal from its dam, usually at 4 to 6 months of age.

weanling. A young horse of either sex, that has been separated from the dam but has not yet reached 1 year old.

yearling. A young horse of either sex from January 1 to December 31 of the year following its birth.

Recommended Reading

Hill, Cherry. *101 Arena Exercises*. North Adams, MA: Storey Publishing, 1995.

———. *101 Longeing and Long Lining Exercises*. Indianapolis: Wiley Publishing Inc., 1998.

———. *The Formative Years: Raising and Training the Young Horse from Birth to Two Years*. Ossining, NY: Breakthrough Publications, 1988.

———. *Cherry Hill's Horse Care for Kids*. North Adams, MA: Storey Publishing, 2002.

———. *Horse for Sale: How to Buy a Horse or Sell the One You Have*. New York: Horsekeeping® Books, 1995.

———. *Horse Handling and Grooming*. North Adams, MA: Storey Publishing, 1990.

———. *Horse Health Care*. North Adams, MA: Storey Publishing, 1997.

———. *Horsekeeping on a Small Acreage,* second edition. North Adams, MA: Storey Publishing, 2005.

———. *Longeing and Long Lining the English and Western Horse*. Indianapolis: Wiley Publishing Inc., 1998.

———. *Making Not Breaking: The First Year Under Saddle*. Ossining, NY: Breakthrough Publications, 1992.

———. *Stablekeeping: A Visual Guide to Safe and Healthy Horsekeeping*. North Adams, MA: Storey Publishing, 2000.

———. *Trailering Your Horse: A Visual Guide to Safe Training and Traveling*. North Adams, MA: Storey Publishing, 2000.

Hill, Cherry, and Richard Klimesh. *Maximum Hoof Power: A Horse Owner's Guide to Shoeing and Soundness*. North Pomfret, VT: Trafalgar Square, 2000.

Kainer, Robert A., and Thomas O. McCracken. *Horse Anatomy: A Coloring Atlas*. Loveland, CO: Alpine Publications Inc., 1998.

Klimesh, Richard, and Cherry Hill. *Horse Housing: How to Plan, Build, and Remodel Barns and Sheds*. North Pomfret, VT: Trafalgar Square, 2002.

Waring, George H. *Horse Behavior,* second edition. Norwich, NY: Noyes Publications, 2003.

Index

Other Storey Titles You Might Enjoy

Among Wild Horses, by Lynne Pomeranz.
An extraordinary photographic journal of three years in the lives of the Pryor
Mountain Mustangs of Montana and Wyoming.
148 pages. Hardcover with jacket. ISBN-13: 978-1-58017-633-0.

The Horse Behavior Problem Solver, by Jessica Jahiel.
A friendly, question-and-answer sourcebook to teach readers how to interpret
problems and develop workable solutions.
352 pages. Paper. ISBN-13: 978-1-58017-524-1.

Horsekeeping on a Small Acreage, by Cherry Hill.
A thoroughly updated, full-color edition of the author's best-selling classic
about how to have efficient operations and happy horses.
320 pages. Paper. ISBN-13: 978-1-58017-535-7.
Hardcover. ISBN-13: 978-1-58017-603-3.

The Horse Training Problem Solver, by Jessica Jahiel.
The third title in a popular series, combining basic training theory, effective
solutions, and handy strategies, in a handy question-and-answer format.
416 pages. Paper. ISBN-13: 978-1-58017-686-6.
Hardcover. ISBN-13:978-1-58017-687-3.

Ride the Right Horse, by Yvonne Barteau.
The key to learning the personality of your horse and working with his strengths.
312 pages. Hardcover with jacket. ISBN-13: 978-1-58017-662-0.

The Rider's Problem Solver, by Jessica Jahiel.
Answers to problems familiar to riders of all levels and styles,
from a clinician and equine behavior expert.
384 pages. Paper. ISBN-13: 978-1-58017-838-9.
Hardcover. ISBN-13: 978-1-58017-839-6.

Storey's Guide to Raising Horses, by Heather Smith Thomas.
The complete guide to intelligent horsekeeping: how to keep a horse
healthy in body and spirit.
512 pages. Paper. ISBN-13: 978-1-58017-127-4.

Storey's Illustrated Guide to 96 Horse Breeds of North America, by Judith Dutson.
A comprehensive encyclopedia filled with full-color photography and in-depth
profiles on the 96 horse breeds that call North America home.
416 pages. Paper. ISBN-13: 978-1-58017-612-5.
Hardcover with jacket. ISBN-13: 978-1-58017-613-2.

These and other books from Storey Publishing are available
wherever quality books are sold or by calling 1-800-441-5700.
Visit us at *www.storey.com*.